STUDIES IN STATISTICS

MAA STUDIES IN MATHEMATICS

Published by
THE MATHEMATICAL ASSOCIATION OF AMERICA

Studies in Mathematics

The Mathematical Association of America

Volume 1: STUDIES IN MODERN ANALYSIS
edited by R. C. Buck

Volume 2: STUDIES IN MODERN ALGEBRA
edited by A. A. Albert

Volume 3: STUDIES IN REAL AND COMPLEX ANALYSIS
edited by I. I. Hirschman, Jr.

Volume 4: STUDIES IN GLOBAL GEOMETRY AND ANALYSIS
edited by S. S. Chern

Volume 5: STUDIES IN MODERN TOPOLOGY
edited by P. J. Hilton

Volume 6: STUDIES IN NUMBER THEORY
edited by W. J. LeVeque

Volume 7: STUDIES IN APPLIED MATHEMATICS
edited by A. H. Taub

Volume 8: STUDIES IN MODEL THEORY
edited by M. D. Morley

Volume 9: STUDIES IN ALGEBRAIC LOGIC
edited by Aubert Daigneault

Volume 10: STUDIES IN OPTIMIZATION
edited by G. B. Dantzig and B. C. Eaves

Volume 11: STUDIES IN GRAPH THEORY, PART I
edited by D. R. Fulkerson

Volume 12: STUDIES IN GRAPH THEORY, PART II
edited by D. R. Fulkerson

Volume 13: STUDIES IN HARMONIC ANALYSIS
edited by J. M. Ash

Volume 14: STUDIES IN ORDINARY DIFFERENTIAL EQUATIONS
edited by Jack Hale

Volume 15: STUDIES IN MATHEMATICAL BIOLOGY, PART I
edited by S. A. Levin

Volume 16: STUDIES IN MATHEMATICAL BIOLOGY, PART II
edited by S. A. Levin

Volume 17: STUDIES IN COMBINATORICS
edited by G.-C. Rota

Volume 18: STUDIES IN PROBABILITY THEORY
edited by Murray Rosenblatt

Volume 19: STUDIES IN STATISTICS
edited by R. V. Hogg

Jack E. Graham
Carleton University

Bruce M. Hill
University of Michigan, Ann Arbor

Robert V. Hogg
University of Iowa

Peter W. M. John
University of Texas, Austin

David S. Moore
Purdue University

Gottfried E. Noether
University of Connecticut, Storrs

J. N. K. Rao
Carleton University

Studies in Mathematics

Volume 19

STUDIES IN STATISTICS

Robert V. Hogg, editor
University of Iowa, Iowa City

Published and distributed by
The Mathematical Association of America

Library of Congress Catalog Card Number 78-71936

Complete Set ISBN 0-88385-100-8
Vol. 19 ISBN 0-88385-119-9

Printed in the United States of America

Current printing (last digit):

10 9 8 7 6 5 4 3 2 1

PREFACE

Statisticians are often appalled at the many misuses of statistical methods, from elementary data analysis, which might be found in newspapers or magazines, to statistics presented in journal articles summarizing experimental research. Frequently, correct statistical summaries do not require a great deal of mathematical training, but rather some good common sense associated with some basic instruction in statistical thinking, namely "making sense out of numbers." Accordingly, many of us in the statistical profession believe that not enough has been done to introduce students (not only those in college, but those in elementary and secondary schools also) to good statistical reasoning.

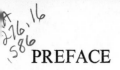

So that I can emphasize the point which I want to make here, let me consider two contrived illustrations, but like ones that occur on a daily basis. The first is about the T (for terrible) disease. Say there are two methods, A and B, of treatment. Suppose 390 patients with this disease were given one of the two treatments: 160 were given A and 230 were given B. The recovery rates were $60/160 = 0.38$ using Treatment A and $65/230 = 0.28$ using Treatment B. While chances of recovering from the T disease are extremely poor and hopefully much more research will be done on it, it seems from these data that it is more favorable to use Treatment A than B. However, by using one classifier, namely sex, the total picture changes as follows.

Sex \ Treatment	*Fraction Recovering* A	B
Male	20/100 = 0.20	50/210 = 0.24
Female	40/60 = 0.67	15/20 = 0.75
Total	60/160 = 0.38	65/230 = 0.28

Now, it is clear from this summary that both men and women would prefer Treatment B, which is different from our earlier conclusion. The lesson here is that you must be extremely careful in reporting averages unless the groups under consideration are fairly homogeneous. Note, in this example, the experiences of men and women were quite different.

To stress this point, let us consider a type of problem that is so often reported in today's newspaper, namely, the difference in the salaries of men and women. Suppose that one company in Iowa City employs 50 men and 50 women and their respective average salaries are $16,000 and $14,000. This seemingly supports the argument that there is discrimination in salaries due to sex. However, by using one classifier, length of employment, that in some sense is a measure of qualification, we obtain the following averages.

Length of employment \ *Sex*	*Average Salaries* Male number	average	Female number	average
less than 5 years	10	10,000	40	12,500
more than 5 years	40	17,400	10	20,000
Totals	50	16,000	50	14,000

From this summary, the women's salaries look better than those of the men. Now, it is not the purpose of this example to argue that women have or have not been discriminated against in the matter of salaries (for the record, it is my personal opinion that there has been some discrimination), but only to emphasize that qualifications (amount of experience, education, and so on) of the individual must be taken into account in arguing that discrimination does or does not exist. Yet frequently, in the newspaper, we find one salary figure representing each of the groups, men and women. With only those two numbers, as in this example, it is impossible to say anything about discrimination in this situation.

Since we feel the need of much more statistical education, the authors welcomed the opportunity to contribute to the *Studies in Statistics*. While, in this volume, we have not discussed elementary problems like those in the illustrations just given, it does provide us the chance of presenting to the members of the broad mathematical community some important aspects of statistics. If through these *Studies* we can only interest a few mathematicians in statistical methods, even if it is from a mathematical point of view, we believe that we will have been successful and helped further statistical education.

As the editor of the *Studies in Statistics*, I was fortunate to be able to find five excellent writing teams, composed of persons well known in their respective research areas. In addition to their chapters, I have written a short introduction that explains much of the terminology needed in these five major papers. Readers with a reasonable background in mathematical statistics could certainly omit my introduction. As a matter of fact, I would urge that each person first read the introductions of the five major chapters. While we have not covered all topics in statistics, this survey will give the reader a good idea about not only the contents of this volume, but about statistical methods in general. Since the chapters are written independently, the reader could then let his or her own interest dictate which materials he or she would like to explore further.

Some would find the chapter by Peter W. M. John very basic. In it, he considers one of the critical areas of statistics, design of experiments, which was first developed formally by that giant of statistics, Sir Ronald A. Fisher. The classical two-sample problem motivates this entire section. John then ties many of the topics of his paper together with a chemical engineering experiment, which involves two different temperatures, two different pressures, and two different acid strengths.

Gottfried Noether then considers many of the topics John does, but from a nonparametric viewpoint. In most nonparametric statistics, ranks of the observations replace the actual observations and then statistics similar to classical (parametric) ones are used. In addition, certain nonparametric procedures result naturally and, in many cases, have efficiencies greater than those of the usual t and F statistics.

The famous chi-square test of Karl Pearson is, in a sense, also nonparametric. David Moore provides an excellent exposition about the interesting mathematical problems that are created with different variations of this statistic. This treatment provides a very modern look at some of the questions about the loss of degrees of freedom that have bothered statisticians from the days of Pearson and Fisher.

A statistical technique about which we hear almost daily is that of the sample survey. Jack Graham and J. N. K. Rao provide a broad summary of the latest tools in this area. There are many interesting questions in mathematical statistics that arise from the theory of surveys. In particular, I find the recent random response technique, in which the surveyor does not know for certain what question the respondent is answering, most intriguing in its use with sensitive questions. Graham and Rao consider this and other recent developments, along with some of the more traditional methods of sample surveys.

Modern theory of decision making will have great appeal to many mathematicians. Bruce Hill has written a very logical approach to the subject, of course, ending with his own favorite topic, Bayesian methods. He does also mention several new research areas that might be of interest to mathematicians, in general, and research statisticians, in particular.

We feel that these *Studies* portray a reasonable picture of statistical methods. While, in such a short volume, we could not tell the whole story, the job will be adequate if at least a few members of the mathematical sciences find it of value. And, personally, I would like to thank first the authors for their excellent cooperation in maintaining a reasonable writing schedule and second Alex Rosenberg who for the Association asked and encouraged me to serve as the editor in the first place.

ROBERT V. HOGG

CONTENTS

AN INTRODUCTION TO MATHEMATICAL STATISTICS

Robert V. Hogg

Random variables are really functions defined on probability spaces and are usually denoted by capital letters such as X, Y, and Z. However, for our purposes, we can think of a random variable as a function of the outcome of a random experiment; that is, X is some measurement that, in a sense, "mathematizes" the outcome. We are frequently interested in the *probability* that X belongs to a certain set A, which is written $\Pr(X \in A)$. Moreover, $X \in A$ is called an *event*. If $F(x) = \Pr(X \leqslant x)$ represents the *(cumulative) distribution function*, then

$$\Pr(X \in A) = \int_A dF(x).$$

If X is a random variable of the continuous type, then

$$\Pr(X \in A) = \int_A f(x)\, dx,$$

where $f(x) = F'(x)$ is the *probability density function* (p.d.f.). That is, this probability is the area underneath the curve $y = f(x)$ and above the set A which is on the x-axis, and it can be thought of as the percentage of the underlying population of x-values in the set A. For random variables of the discrete type, summations replace integrations; specific mention of this fact will not be made again. That is, only continuous-type random variables are considered in this introduction, and either $F(x)$ or $f(x)$ describes the distribution of probability associated with X.

Let $u(X)$ be a function X and hence a random variable itself. The *mathematical expectation* (or *expected value*) of $u(X)$ is

$$E[u(X)] = \int_{-\infty}^{\infty} u(x)\, dF(x) = \int_{-\infty}^{\infty} u(x) f(x)\, dx.$$

Two important characteristics of the distribution of X are the *mean* $\mu = E(X)$ and the *variance* $\sigma^2 = E[(X - \mu)^2]$, which is sometimes denoted by $\text{var}(X)$ or $V(X)$. The positive square root σ of the variance is called the *standard deviation* of X. In the physical sciences, μ and σ could be thought of as the centroid and the radius of gyration of a body of weights (probabilities). The *characteristic function* of X, namely

$$\varphi(t) = E(e^{itX}) = \int_{-\infty}^{\infty} e^{itx} f(x)\, dx$$

is essentially the Fourier transform of $f(x)$. From transform theory, knowledge of $\varphi(t)$ is equivalent to that of $f(x)$ or $F(x)$, a fact that proves extremely useful in distribution theory.

The *normal* and *chi-square* distributions are two important ones in statistics. The former has that bell-shaped p.d.f.

$$f(x) = \frac{1}{\sqrt{2\pi}\sigma} \exp\left[-\frac{(x-\mu)^2}{2\sigma^2}\right], \qquad -\infty < x < \infty.$$

Here the parameters μ and σ do represent the mean and standard deviation, respectively. Sometimes we say that a random variable with this p.d.f. is $N(\mu, \sigma^2)$. The distribution $N(0, 1)$ is called the standard normal distribution. The chi-square distribution has p.d.f.

$$f(x) = \frac{1}{\Gamma(r/2)2^{r/2}} x^{r/2-1}e^{-x/2}, \qquad 0 < x < \infty,$$

zero elsewhere, for which the abbreviation is $\chi^2(r)$. The parameter r is a positive integer called *degrees of freedom* (d.f.) and here $\mu = r$ and $\sigma^2 = 2r$.

Suppose we have two or more random variables. For illustration, say X and Y have the *joint p.d.f.* $f(x, y)$, then

$$\Pr[(X, Y) \in A] = \int_A \int f(x, y)\, dx\, dy,$$

where A is a set in the xy-plane. The total volume under $z = f(x, y)$ and above the xy-plane is equal to one. The *expected value* of $u(X, Y)$ is given by

$$E[u(X, Y)] = \int_{-\infty}^{\infty} \int_{-\infty}^{\infty} u(x, y) f(x, y)\, dx\, dy.$$

Important characteristics of the joint distribution are the means $\mu_X = E(X)$ and $\mu_Y = E(Y)$, the variances $\sigma_X^2 = \text{var}(X) = E[(X - \mu_X)^2]$ and $\sigma_Y^2 = \text{var}(Y) = E[(Y - \mu_Y)^2]$, and the *covariance*

$$\sigma_{XY} = \text{cov}(X, Y) = E[(X - \mu_X)(Y - \mu_Y)].$$

The *correlation coefficient* of X and Y is $\rho_{XY} = \sigma_{XY}/(\sigma_X \sigma_Y)$. The matrix

$$\begin{pmatrix} \sigma_X^2 & \sigma_{XY} \\ \sigma_{XY} & \sigma_Y^2 \end{pmatrix}$$

is frequently called the *variance-covariance* matrix. An important p-variate distribution of $\mathbf{X}$, where $\mathbf{X}' = (X_1, X_2, \ldots, X_p)$, is the *normal multivariate* one with p.d.f.

$$\frac{1}{(2\pi)^{p/2}\sqrt{|\boldsymbol{\Sigma}|}} \exp\left[-\frac{(\mathbf{x} - \boldsymbol{\mu})'\boldsymbol{\Sigma}^{-1}(\mathbf{x} - \boldsymbol{\mu})}{2}\right],$$

where $\boldsymbol{\mu}$ and $\boldsymbol{\Sigma}$ are the matrices of the means and the variances and covariances. It is denoted by $N_p(\boldsymbol{\mu}, \boldsymbol{\Sigma})$.

The functions

$$f_1(x) = \int_{-\infty}^{\infty} f(x, y)\, dy, \qquad f_2(y) = \int_{-\infty}^{\infty} f(x, y)\, dx$$

are the *marginal p.d.f.'s* of X and Y, respectively. The *conditional p.d.f.'s*

$$f_1(x|y) = \frac{f(x, y)}{f_2(y)}, \qquad f_2(y|x) = \frac{f(x, y)}{f_1(x)}$$

are those of X, given $Y = y$, and of Y, given $X = x$. In the p-variate normal case, each marginal p.d.f. is normal with the appropriate mean and variance, and the conditional p.d.f. of X_i, given $X_j = x_j, j \neq i$, is also normal.

If $f(x, y, z)$ is the joint p.d.f. of X, Y, and Z, we say that these three variables are *mutually independent* if and only if

$$f(x, y, z) = f_1(x)f_2(y)f_3(z),$$

where $f_1(x)$, $f_2(y)$, and $f_3(z)$ are the respective marginal p.d.f.'s for X, Y, and Z. If, in addition, the three marginal distributions are the same, then the three random variables are *independent and identically distributed* (i.i.d.).

A fundamental idea needed in statistics, but one which is difficult for many students, is that of a distribution of a function of one, or two, or more random variables. Theoretically, it is handled extremely easily. Suppose $Z = u(X, Y)$ is a function of two random variables X and Y with p.d.f. $f(x, y)$, then the distribution function of Z is

$$G(z) = \Pr(Z \leqslant z) = \int_A\!\!\int f(x, y)\, dx\, dy,$$

where the set A in the xy-plane is described by $u(x, y) \leqslant z$. (It should be clear that here and elsewhere, we are discussing Borel measurable functions.) Actually finding $G(z)$, or the corresponding p.d.f. $g(z) = G'(z)$, in a number of special cases, is a rather major effort in a course in mathematical statistics. Three important examples are:

(a) If X and Y are independent with distributions $N(0, 1)$ and $\chi^2(r)$, respectively, then $T = X/\sqrt{Y/r}$ has a *Student t-distribution* with r d.f.

(b) If X and Y are independent with distributions $\chi^2(r_1)$ and $\chi^2(r_2)$, then $F = (X/r_1)/(Y/r_2)$ has an *F distribution* with r_1 and r_2 d.f.

(c) If $\mathbf{X}$ has a p-variate normal distribution $N_p(\boldsymbol{\mu}, \boldsymbol{\Sigma})$, then $\mathbf{c}'\mathbf{X}$ is $N(\mathbf{c}'\boldsymbol{\mu}, \mathbf{c}'\boldsymbol{\Sigma}\mathbf{c})$, where $\mathbf{c}$ is a vector of constants, and $(\mathbf{X} - \boldsymbol{\mu})'\boldsymbol{\Sigma}^{-1}(\mathbf{X} - \boldsymbol{\mu})$ is $\chi^2(p)$.

If $X_1, X_2, \ldots, X_n$ are i.i.d., we sometimes say that they are *items of a random sample* from a distribution with p.d.f. $f(x)$, which is the common marginal one of all n variables. Let $x_1, x_2, \ldots, x_n$ be the observed values of the items $X_1, X_2, \ldots, X_n$ of a random sample. By placing the weight $1/n$ on each x_i, we obtain a distribution of probability. The corresponding distribution function $F_n(x)$ is called the *empirical distribution function* and it is of the discrete type. The mean and the variance of this discrete distribution are

$$\bar{x} = \left(\sum_{i=1}^{n} x_i\right)\Big/ n, \qquad s^2 = \sum_{i=1}^{n} (x_i - \bar{x})^2/n,$$

respectively, and are often called the mean and the variance of the observed sample. Since, with the corresponding random variables, $\bar{X}$ and S^2, we have that

$$E(\bar{X}) = \mu, \qquad E(S^2) = \frac{n-1}{n}\sigma^2,$$

we call $\bar{X}$ an *unbiased estimator* of μ while S^2 is a *biased estimator* of σ^2. For the purpose of obtaining an unbiased estimator of σ^2, many authors define

$$\frac{nS^2}{n-1} = \frac{1}{n-1}\sum_{i=1}^{n} (X_i - \bar{X})^2 = S_1^2,$$

as the variance of the sample.

More generally, if we have a *statistic* (a function of the sample items), say $Z = u(X_1, X_2, \ldots, X_n)$, then Z would be a good estimator of a parameter θ provided the observed Z, say $z = u(x_1, x_2, \ldots, x_n)$, is close to θ. To accomplish this frequently, we would want Z to have small (or zero) bias and small variance. Two ways of doing this would be to require that

(i) Z be *unbiased with minimum variance* (UMV) among all unbiased estimators of θ, or
(ii) Z be that statistic which minimizes $E[(Z - \theta)^2]$, that is, the *minimum mean-square-error* estimator of θ.

Another good principle of point estimation which appeals to our intuition is that of *maximum likelihood.* Say the sample arises from a distribution with p.d.f. $f(x|\theta)$, where θ is unknown. The joint p.d.f. of $X_1, X_2, \ldots, X_n$ thought of as a function of θ is

$$L(\theta) = f(x_1|\theta)f(x_2|\theta)\cdots f(X_n|\theta),$$

and is called the likelihood function. The value of θ that maximizes $L(\theta)$, the joint p.d.f., is called the maximum likelihood estimator of θ and is often denoted by $\hat{\boldsymbol{\theta}}$. For the normal distribution $N(\mu, \sigma^2)$, $\hat{\mu} = \bar{X}$ and $\hat{\sigma}^2 = S^2$.

Usually statisticians provide more than a point estimate of a parameter θ. That is, along with a point estimate, say $\hat{\theta}$, they would give some notion of the error structure of this estimate. For illustration, in the normal case, $\hat{\mu} = \bar{X}$, but it is also true that $T = \sqrt{n}(\bar{X} - \mu)/S_1$ has a t-distribution with $n - 1$ d.f. If t^* is selected so that $\Pr(|T| \leqslant t^*) = 0.95$, say, then the interval described by the endpoints $\bar{X} \pm t^*S_1/\sqrt{n}$ has the probability 0.95 of including the unknown mean μ because $|T| \leqslant t^*$ is equivalent to $\bar{X} - t^*S_1/\sqrt{n} \leqslant \mu \leqslant \bar{X} + t^*S_1/\sqrt{n}$. Thus the observed interval $\bar{x} \pm t^*s_1/\sqrt{n}$ provides a 95% *confidence interval* for μ. That is, the $t^*s_1/\sqrt{n}$ (or $s_1/\sqrt{n}$ alone) provides a measure of the error structure of the estimator $\bar{X}$.

Estimation is one major aspect of *statistical inference.* Another is concerned with tests of statistical hypotheses. A *statistical hypothesis* is a statement about a distribution of probability, for example, $H_0 : \mu = \mu_0$. A *test* of H_0 is a procedure, based on a random sample, by which we would accept or reject H_0. The thought that we might reject H_0 means that we must have in mind some *alternative hypothesis*, say $H_1 : \mu = \mu_1 > \mu_0$. In our normal example, we might accept $H_0 : \mu = \mu_0$ or $H_1 : \mu = \mu_1$ depending on whether the observed sample mean $\bar{x}$ is closer to μ_0 or μ_1. In such a procedure, we clearly can make one of two types of errors: *Type-I error* is rejecting H_0 when H_0 is true and *Type-II error* is accepting H_0 when H_0 is false. The probabilities of these two types of errors are denoted by α and β, respectively, and α is frequently called the *significance level* of the test. The *power of the test* at μ_1 is $1 - \beta$.

To return to our normal case, say we wished to test the statistical hypothesis $H_0 : \mu = \mu_0$ against the *one-sided* alternative hypothesis $H_1 : \mu > \mu_0$. Consider the statistic $T = \sqrt{n}(\bar{X} - \mu_0)/S_1$, which has a t-distribution with $n - 1$ d.f. provided H_0 is true. Find a *critical value* t^* so that $\Pr(T \geqslant t^* | H_0) = \alpha$. If we reject H_0 and accept H_1 if the observed $T \geqslant t^*$, then the significance level of this test is α.

Clearly, there are many other topics in mathematical statistics; but, hopefully, with this brief introduction the reader will be able to understand much of the discussion in the five major chapters that follow.

EXPERIMENTAL DESIGN

Peter W. M. John

1. INTRODUCTION

The design of experiments is concerned with planning experimental programs in the most efficient way. It entails applying a basic set of mathematical principles to reach decisions about the allocation of the experimental resources. It is a branch of applied mathematics and of applied statistics. Some of its aspects are little more than basic common sense; others are almost exotic excursions into the worlds of finite projective planes, Galois fields, and finite geometries.

The mathematical prerequisites are not extensive. We do not need any measure theory or complex analysis. The main mathematical tools are linear algebra, Abelian groups, and some n-dimensional geometry. From statistics we call upon the standard results for the normal distribution: some of them are stated without proof in this article, but a knowledge of their derivation is not essential at this stage, and the reader who is prepared to accept them at face value will get along very well.

Some scientists are surprised by the suggestion that a mathematician can make a useful contribution to an experiment except to carry out computations. Indeed, so are some mathematicians. In the age of computers, they no longer need us to help with the arithmetic, and few of us can claim much expertise in the many and varied areas of science in which experiments are performed. But there is more to data analysis than merely feeding one's data to a computer. With a modern computer we can fit a straight line to almost any stack of data; we can also fit quadratics and sine curves, planes and splines, but all these avail us little if the data is a hodgepodge to begin with, and can even be deceptive.

Anyone who has had much experience of analyzing the data from other people's experiments can tell sad stories of thousands of dollars being wasted on experiments that were so poorly planned that, after the data were collected, there was no way in which it could give a valid answer to the scientist's original question. He can also tell of experiments in which a little skilled advice in the planning stage would have led either to more information for the same amount of effort or to a smaller experiment and a saving of money and time.

If the data is to be analyzed by the use of statistical procedures, the scientist should consider carefully before the experiment is carried out what form of statistical analysis will be appropriate and whether or not his data is going to be amenable to it. Therefore, he should consult the statistician at the planning stage rather than wait until after the data has been collected.

Between them—and we emphasize cooperation—they can address the key issues: Will it be possible to answer the questions he is posing with the resources available to him? What is the most efficient and economical experimental program to attain his objectives? Can the experiment be designed so that the data will provide answers to more questions at the same time?

This article is concerned with the planning and design of experiments, which as we have seen includes the analysis, and in particular with the underlying mathematical ideas. The fundamental concepts are those of the linear model and the analysis of variance.

The design of experiments as we know it began with the work of Sir Ronald A. Fisher between the two World Wars. He had taken a degree in mathematics at Cambridge University before turning to statistics and genetics. As the director of statistics at the Rothamsted agricultural research station in Great Britain, he was involved in field and laboratory experiments. He pioneered the analysis of variance, which is the basis of design of experiments. He wrote two books on the subject, *Statistical Methods for Research Workers* (first edition 1925), and *The Design of Experiments*; the latter was first published in 1935, and the eighth (posthumous) edition appeared in 1966. The treatment is essentially nonmathematical, and the mathematically inclined reader will find them heavy going. Mathematicians may prefer to sample the more modern books by Kempthorne (1952), Cox (1958), Scheffé (1959), and John (1971).

We detour briefly to refresh the memories of some of our readers about Student's t statistic. "Student" was the pen name of William Sealy Gosset, a chemist who came from the Guinness brewery in Dublin to study with Karl Pearson at the Galton Laboratory in London at the beginning of the century. He was probably the first industrial statistician, and the problem on which he worked is basic to applied statistics: how do we modify the normal test when we have to estimate the variance σ^2 from the data?

If $\bar{x}$ is the mean of n independent observations, each of which is normally distributed with mean μ and variance σ^2, then $\bar{x}$ is itself normally distributed with mean μ and variance σ^2/n, and so $z = (\bar{x} - \mu)\sqrt{n}/\sigma$ has a standard normal distribution: $E(z) = 0$, $V(z) = 1$. It follows that the probability that $|z| > 1.96$ is 0.05. The classical test procedure for the hypothesis $H_0 : \mu = 0$ is to reject the hypothesis if $|z| > 1.96$; this procedure has an error risk $\alpha = 5\%$. If we do not know σ^2, we estimate it from the data by $s^2 = \sum (x_i - \bar{x})^2/(n - 1)$, which is the quadratic estimate with $\phi = n - 1$ degrees of freedom. The test statistic for $H_0 : \mu = 0$ now becomes $t = \bar{x}\sqrt{n}/s$, and we reject H_0 if $|t| > t^*_{\alpha,\phi}$ where $t^*_{\alpha,\phi}$ is the tabulated value corresponding to ϕ and our chosen α. With $\alpha = 0.05$, $t^*_{\alpha,\phi}$ will be slightly larger than 1.96, approaching that value as $\phi \to \infty$. If we have independent samples $(x_1, \ldots, x_n)$, $(y_1, \ldots, y_n)$ of size n from two normal populations with means μ_1, μ_2 and common variance σ^2, and we wish to test the hypothesis $H_0: \mu_1 = \mu_2$, the appropriate t statistic is $t =$

$(\bar{x} - \bar{y})/\sqrt{(2s^2/n)}$, where $s^2 = [\sum (x_i - \bar{x})^2 + \sum (y_i - \bar{y})^2]/(2n - 2)$; t has $2(n - 1)$ d.f.

We have just referred to one of the simplest experiments, comparing the means of two populations, which we assume to be normal. It is clear that the thing to do is to take some observations from each population and then to look at $\bar{x} - \bar{y} = d$. If d is "large," we conclude that $\mu_1 \neq \mu_2$, but how big must d become to be classified as "large"? That depends upon σ^2, and it is important to realize that we turn to the experiment itself to provide not only d, but also a valid estimate of the error which is used in the denominator of the test statistic.

In the sections which follow we shall give short discussions of some of the many topics that are embraced by the subject of experimental design. They will be tied together somewhat loosely by the thread of a chemical engineering experiment. Suppose that an engineer has to experiment on a pilot plant by running the process at two different temperatures, two different pressures, and two different acid strengths, $2 \times 2 \times 2 = 8$ sets of operating conditions in all, but that he is handicapped by being able to make only three runs on each batch of raw material. What should he do?

We could as well have drawn our connecting example from agriculture, biochemistry, pharmacy, or veterinary medicine, to mention but a few areas. By and large, the principles of our experimental design are the same. Just as we can apply Student's t test to data on sheep, potatoes, students, or automobiles, so we can take a larger experimental situation and express it in the abstract form of the linear model. This ability to see the essential unity of the field distinguishes the mathematician from scientists who tend to view problems only in terms of their own field of specialization. It is the product of his mathematical training with its emphasis on abstract thinking.

2. MULTIPLE REGRESSION

The idea of the linear model is fundamental to the design and analysis of experiments. Let us picture a chemical engineer carrying out an industrial experiment in which n runs are made on a process in a pilot plant at various temperatures and pressures, and the yield of each run is observed. We denote the yield on the ith run by y_i, and the temperature and pressure at which the run was made by x_{i1} and x_{i2}, respectively. The sample space is two-dimensional Euclidean space with x_1 denoting the temperature and x_2 the pressure. The model that we assume sees the yield, y, as the sum of four terms: a constant term, a first degree contribution for the temperature, a similar contribution for the pressure, and, finally, a random component described as error or noise. We have

$$y_i = \beta_0 + \beta_1 x_{i1} + \beta_2 x_{i2} + e_i.$$

More generally, we can consider p factors $x_1, \ldots, x_p$ with corresponding coefficients $\beta_1, \ldots, \beta_p$. The coefficients $\beta_0, \beta_1, \ldots$ are unknown constants. The errors e_i are assumed to be independent normal random variables with $E(e_i) = 0$ and $V(e_i) = \sigma^2$.

The sample points (x_1, x_2) may have been chosen in some systematic fashion, as would be the case with a designed experiment, or we may be given a set of data in which the sample points were taken virtually at random. In either case, the usual objective of the analysis is to obtain estimates of the coefficients β_i, which we may use to predict the yield of a future run on the same pilot plant.

We introduce a dummy variable $x_0 \equiv 1$ to go with β_0 so as to make the model for $E(y)$ homogeneous, and write the model in matrix form as

$$\mathbf{Y} = \mathbf{X}\boldsymbol{\beta} + \mathbf{e},$$

where $\mathbf{Y}$ is the vector of observed yields, having n elements; $\mathbf{X}$ is a matrix of n rows and $p + 1$ columns, the first of which is a vector of unit elements denoted by $\mathbf{1}$; $\mathbf{e}$ is the vector of random errors. Then $\mathbf{e} \sim N(\mathbf{0}, \mathbf{I}\sigma^2)$ and $\mathbf{Y} \sim N(\mathbf{X}\boldsymbol{\beta}, \mathbf{I}\sigma^2)$.

The method of least squares consists in taking as the estimates of the coefficients $\boldsymbol{\beta}$ the set of values, $\hat{\boldsymbol{\beta}}$, which minimize the sum of the squares of the residuals between the observed values $\mathbf{Y}$ of the yield and the estimated values. These least squares estimates, $\hat{\mathbf{Y}} = \mathbf{X}\hat{\boldsymbol{\beta}}$, provide the minimum $S_e = \sum_{i=1}^{n} (y_i - \hat{y}_i)^2 = (\mathbf{Y} - \hat{\mathbf{Y}})' \times$

$(\mathbf{Y} - \hat{\mathbf{Y}}) = (\mathbf{Y} - \mathbf{X}\hat{\boldsymbol{\beta}})'(\mathbf{Y} - \mathbf{X}\hat{\boldsymbol{\beta}})$, where $\hat{\boldsymbol{\beta}}$ is a solution vector to the set of normal equations

$$\mathbf{X}'\mathbf{Y} = \mathbf{X}'\mathbf{X}\hat{\boldsymbol{\beta}}.$$

S_e is called either the sum of squares for error or the residual sum of squares.

There are two cases to be considered, that in which $(\mathbf{X}'\mathbf{X})^{-1}$ does exist, and that in which it does not exist. The latter case occurs in the standard experimental designs.

If $(\mathbf{X}'\mathbf{X})^{-1}$ exists, the vector of estimates is $\hat{\boldsymbol{\beta}} = (\mathbf{X}'\mathbf{X})^{-1}\mathbf{X}'\mathbf{Y}$, and it is readily shown that $E(\hat{\boldsymbol{\beta}}) = \boldsymbol{\beta}$ and that the variance-covariance matrix of the estimates is $\text{cov}(\hat{\boldsymbol{\beta}}) = (\mathbf{X}'\mathbf{X})^{-1}\sigma^2$. Furthermore, the unbiased quadratic estimate of σ^2 is $s^2 = S_e/(n - p - 1)$. We may test the hypothesis $H_0 : \beta_i = 0$ by a t test, using for the variance of $\hat{\beta}_i$ in the denominator of the t statistic s^2 multiplied by the ith diagonal element of $(\mathbf{X}'\mathbf{X})^{-1}$.

We may also test the hypothesis that a particular subset of the regression coefficients are simultaneously zero. Suppose that we divide $\boldsymbol{\beta}$ into two subvectors, $\boldsymbol{\beta}' = (\boldsymbol{\beta}_1', \boldsymbol{\beta}_2')$, where $\boldsymbol{\beta}_2$ has q elements and $\boldsymbol{\beta}_1$ has $p + 1 - q$, and that we wish to test the hypothesis $\boldsymbol{\beta}_2 = \mathbf{0}$. We can fit in turn to the data two linear models: the "long" model, $E(y) = \sum_{i=0}^{p} \beta_i x_i$, and the "short" model, $E(y) = \sum_{i=0}^{p-q} \beta_i x_i$, or $E(\mathbf{Y}) = \mathbf{X}_1\boldsymbol{\beta}_1 + \mathbf{X}_2\boldsymbol{\beta}_2$ and $E(\mathbf{Y}) = \mathbf{X}_1\boldsymbol{\beta}_1$, respectively, where $\mathbf{X}$ has been partitioned into $(\mathbf{X}_1, \mathbf{X}_2)$. Suppose that the sums of squares for error with the two models are S_e and S_e', respectively; S_e' will be at least as large as S_e.

We may argue that $s^2 = S_e/(n - p - 1)$ is an unbiased estimate of σ^2 and that, under the hypothesis $\boldsymbol{\beta}_2 = \mathbf{0}$, S_e' is an unbiased estimate of $(n - p - 1 + q)\sigma^2$, so that $E(S_e' - S_e) = q\sigma^2$. Thus, if the null hypothesis is true, $(S_e' - S_e)/q$ and s^2 both estimate σ^2 and so their ratio $\mathscr{F}$ should be close to unity. However, if the null hypothesis is false, suppressing $\boldsymbol{\beta}_2$ from the model should give a disproportionately large increase in the error sum of squares and $\mathscr{F}$ should exceed unity. The ratio $\mathscr{F}$ provides a test statistic with the upper tail of the distribution as the critical region.

This argument does not depend upon normality. With normality, we may argue more formally: S_e/σ^2 has a χ^2 distribution with

$n - p - 1$ d.f.; under the hypothesis $\boldsymbol{\beta}_2 = \mathbf{0}$, $(S_e' - S_e)/\sigma^2$ has a χ^2 distribution with q d.f. The two quadratic forms are independent, and so the ratio of their mean squares, $\mathscr{F}$, has a $F(q, n - p - 1)$ distribution. In the case where $\boldsymbol{\beta}_2$ consists of a single coefficient, we have $\mathscr{F} = t^2$ and this test is equivalent to the t test mentioned earlier.

We may indeed consider a sequence of nested models, passing from one to the next shorter model by dropping a few more coefficients, and obtain a subdivision of $\mathbf{Y'Y}$ into component sums of squares

$$\mathbf{Y'Y} = \mathbf{Y'A}_0\mathbf{Y} + \mathbf{Y'A}_1\mathbf{Y} + \mathbf{Y'A}_2\mathbf{Y} + \cdots,$$

where $\mathbf{Y'A}_0\mathbf{Y} = n^{-1}\mathbf{Y'11'Y} = n(\bar{y})^2$ and $\mathbf{Y'A}_i\mathbf{Y}$ is the increase in the error sum of squares as we pass from one model to the next. Each of these quadratic forms divided by σ^2 will have a central, or perhaps noncentral, χ^2 distribution, and they are mutually independent. The fundamental theorem of Cochran states that a necessary and sufficient condition for the forms $\mathbf{Y'A}_i\mathbf{Y}/\sigma^2$ to have independent χ^2 distributions with the number of degrees of freedom equal to the rank of $\mathbf{A}_i$ is that any one of the following three equivalent conditions shall hold:

(i) each $\mathbf{A}_i$ is idempotent;
(ii) $\mathbf{A}_h\mathbf{A}_i = \mathbf{0}$ for all pairs h, i $(h \neq i)$;
(iii) the sum of the ranks of the $\mathbf{A}_i$ is n.

Nothing that has been said implies that the x-variables, or predictor variables, have to be independent. They may be functionally related. If our engineer should decide that a quadratic model would be more appropriate than the first degree model that we proposed above, he could fit

$$y = \beta_0 + \beta_1 x_1 + \beta_2 x_2 + \beta_{11} x_1^2 + \beta_{12} x_1 x_2 + \beta_{22} x_2^2 + e$$

by setting $x_1^2 = x_3$, $x_1 x_2 = x_4$ and $x_2^2 = x_5$.

This is the least squares procedure that was introduced by Gauss. There are two strong arguments in its favor. Under normality, the least squares estimates are the maximum likelihood estimates. The estimates are linear functions of the observations or, for short, linear estimates. The Gauss-Markov theorem does not depend upon normality, merely upon the independence and equality of variances of the e_i; it states that in the class of all unbiased linear estimates of the regression coefficients, the least squares estimates are the unique estimates with minimum variance.

We have mentioned the possibility that $\mathbf{X}$ could have less than full rank; an example is given in the next section. The normal equations remain the same and are consistent; the difficulty lies in the fact that there is an infinite number of solution vectors $\tilde{\boldsymbol{\beta}}$. There are two ways of handling this problem. One, which we shall not discuss, is to reduce the number of parameters to equal the rank of $\mathbf{X}$ by reparameterizing. The other is to impose enough side conditions upon the parameters to obtain a unique solution vector. It is appropriate to ask how far we can go in imposing side conditions without disastrous results. Different choices of side conditions give different solution vectors $\tilde{\boldsymbol{\beta}}$.

R. C. Bose introduced the idea of estimable functions. What linear combinations of the parameters can we estimate from our data? We can certainly estimate $E(y_i)$; y_i itself is an unbiased estimate of its expectation, although we may be able to find a better one; similarly, we can plausibly estimate $E(y_1) + 2E(y_2)$ by $y_1 + 2y_2$. Let $\psi = \boldsymbol{\lambda}'\boldsymbol{\beta}$ be a linear combination of the parameters. We say that ψ is estimable if, and only if, there exists a linear combination $\mathbf{c}'\mathbf{Y}$ of the observations such that $E(\mathbf{c}'\mathbf{Y}) = \psi$. This is equivalent to saying that $\boldsymbol{\lambda}'$ is a linear combination of the rows of $\mathbf{X}$.

Suppose that the rank of $\mathbf{X}$ is $p + 1 - k$. The space of estimable functions is a $(p + 1 - k)$-dimensional vector space. We impose k side conditions upon the parameters in the form $\mathbf{H}\boldsymbol{\beta} = \mathbf{0}$, where $\mathbf{H}$ has k rows and $p + 1$ columns, in such a way that $(\mathbf{X}', \mathbf{H}')$ has rank $(p + 1)$. This implies that the rows of $\mathbf{H}$ are linearly independent of the rows of $\mathbf{X}$ and of one another and that the functions $\mathbf{H}\boldsymbol{\beta}$ that we arbitrarily set to zero are not estimable functions. Substituting $(\mathbf{Y}', \mathbf{0}')'$ for $\mathbf{Y}$ and $(\mathbf{X}', \mathbf{H}')'$ for $\mathbf{X}$, we have a modified set of normal equations of full rank

$$(\mathbf{X}', \mathbf{H}')(\mathbf{Y}', \mathbf{0}')' = \mathbf{X}'\mathbf{Y} = (\mathbf{X}'\mathbf{X} + \mathbf{H}'\mathbf{H})\tilde{\boldsymbol{\beta}}.$$

The estimate of an estimable function ψ is $\hat{\psi} = \boldsymbol{\lambda}'\tilde{\boldsymbol{\beta}} =$

$\boldsymbol{\lambda}'(\mathbf{X}'\mathbf{X} + \mathbf{H}'\mathbf{H})^{-1}\mathbf{X}'\mathbf{Y}$; it is unique in the sense that no matter what (nonestimable) side conditions we use the value of $\hat{\psi}$ is unchanged. We have the following more general result. Let $\mathbf{PX}'\mathbf{Y} = \tilde{\boldsymbol{\beta}}$ be any solution vector to the normal equations, and let ψ be an estimable function. Then $\hat{\psi} = \boldsymbol{\lambda}'\mathbf{PX}'\mathbf{Y}$ is the unique unbiased linear estimate of ψ with minimum variance, namely $V(\hat{\psi}) = \boldsymbol{\lambda}'\mathbf{P}\boldsymbol{\lambda}\sigma^2$.

3. THE ONE-WAY LAYOUT

The early development of analysis of variance and the design of experiments was in agricultural research, so the terminology of the subject reflects its agricultural ancestry. In the simplest experiment, an agronomist divides a field into plots on which he sows different varieties of a crop, such as wheat, with one variety to a plot. In the summer he harvests his plots and compares the yields of the several varieties. Alternatively, he might plant only one variety of wheat, but apply different "treatments" such as different levels of fertilizer.

Suppose that he has v varieties of wheat and r plots for each variety with a total of $n = rv$ plots. He assigns the plots to the varieties by some random procedure, sows, and waits. Let y_{ij} denote the yield in the jth of the plots that was given the ith variety, $i = 1, \ldots, v$ and $j = 1, \ldots, r$. We take the following linear model for the yield:

$$y_{ij} = \mu + \tau_i + e_{ij},$$

where μ represents an overall average yield for all the varieties, τ_i represents the difference in yield between the ith variety and that overall average, and e_{ij} is again a random noise term, such that the e_{ij} are independently normally distributed with zero expectation and variance σ^2. When we write the model as

$$y_{ij} = \mu x_0 + \sum_{i=1}^{v} \tau_i x_i + e_{ij},$$

with $x_0 \equiv 1$ and $x_i = 1$ if the observation is on the ith variety, and zero otherwise, we have a model that follows the pattern of the previous section with one important modification.

If we order the observations by varieties, so that the first r elements of the observation vector **Y** are the observations on the first variety, and so on, **X**, which now has $v + 1$ columns which we number zero through v, has a simple structure. The first r rows have unit elements in the zero and first columns and zeros elsewhere. The ith set of r rows has unities in the zero and ith columns. There are only v distinct rows; **X** and **X'X** have rank v with $v + 1$ columns; $(\mathbf{X}'\mathbf{X})^{-1}$ does not exist.

The normal equations for the estimates of μ and $\boldsymbol{\tau} = (\tau_1, \ldots, \tau_v)'$ have the same form as in the previous section; $\sum x_0 y_i$ becomes G, the grand total, and $\sum x_i y_i$ is the total, T_i, of the observations on the ith variety, and the equations simplify to

$$G = n\hat{\mu} + r \sum_i \hat{\tau}_i, \qquad T_i = r\hat{\mu} + r\hat{\tau}_i.$$

The usual procedure is to solve the equations under the side condition $\sum \tau_i = \mathbf{1}'\boldsymbol{\tau} = 0$, in which case we obtain the obvious estimates:

$$\hat{\mu} = G/n = \bar{y}, \qquad \hat{\tau}_i = T_i/r - \hat{\mu} = y_{i.} - \bar{y},$$

where $y_{i.} = T_i/r$.

There are usually two broad questions with which the agronomist may be concerned. Do the varieties differ with respect to yield? If so, which varieties differ, and by how much? The first question can be framed in terms of the hypothesis $H_0 : \boldsymbol{\tau} = \mathbf{0}$, under which our model becomes $y_{ij} = \mu + e_{ij}$.

With the "long" model, containing $\boldsymbol{\tau}$, the sum of squares for residuals is $S_e = \sum_i \sum_j (y_{ij} - y_{i\cdot})^2$, which is merely the pooled sums of squares of deviations for the several samples. Under the null hypothesis we have $S_e' = \sum_i \sum_j (y_{ij} - \bar{y})^2$. We denote $S_e' - S_e$ by S_t and call it the sum of squares for varieties (or treatments). Then, $S_t = r \sum (y_{i.} - \bar{y})^2 = \sum T_i^2/r - G^2/n$. Under either model S_e/σ^2 has a χ^2 distribution with $n - v = v(r - 1)$ d.f.; under the null hypothesis S_t/σ^2 has an independent χ^2 distribution with $v - 1$ d.f., and we are led to the F statistic: $\mathscr{F} = [S_t/(v - 1)]/s^2$, where $s^2 = S_e/(n - v)$. It is interesting to note that the generalized likelihood ratio procedure leads to the same test for H_0.

Neither μ nor τ_i is estimable, but the estimable functions include all contrasts in the varieties, by which we mean all functions in the class $\psi = \mathbf{c}'\boldsymbol{\tau}$, where $\mathbf{c}'\mathbf{1} = \sum c_i = 0$. We shall call the simplest contrasts, $\tau_h - \tau_i$, comparisons. We note that $\hat{\psi} = \sum c_i y_{i\cdot}$ and $V(\hat{\psi}) = r^{-1}\mathbf{c}'\mathbf{c}\sigma^2$. A simple procedure for deciding which varieties differ is based upon the two-sample t-test. We declare that the hth and ith varieties differ in respect to yield, if, and only if, $|y_{h\cdot} - y_{i\cdot}| > t^* s\sqrt{2/r}$, where t^* is the critical value of Student's t for $(n - v)$ d.f. and the appropriate level of α. The quantity on the right side of the inequality is called Fisher's LSD (least significant difference). Objections have been raised to this simple procedure on the grounds that when we apply the LSD to all $v(v - 1)/2$ comparisons, the probability that we shall make at least one type one error will exceed α. For instance, we shall be looking at the difference between the largest and smallest of the sample means, which is, under the null hypothesis, the range of a sample of v normal variables with variance σ^2/r. This topic of multiple comparisons would take us beyond the scope of this article. It will suffice to say that Tukey introduced an LSD based upon the Studentized range of v variates; Newman and Keuls and, later, Duncan have significant differences that fall some way between Tukey's and Fisher's, while Scheffé's S statistic faces the problem for general contrasts. These methods are expounded in the book by Miller (1966).

4. BLOCKING

If there is considerable variation in fertility over the whole planting area, the agronomist may decide to divide his field first into strips, called blocks, and then to subdivide the blocks into plots. Having done this, he would, if possible, give each variety one plot in each block with the idea that when we estimate a comparison by $y_{h\cdot} - y_{i\cdot}$ we shall in each mean be averaging over all the blocks, so that when we perform the subtraction the differences in fertility from block to block will cancel out. Our chemical engineer in his pilot plant may choose to consider a run under a set of experimental conditions as a "plot" and the set of runs made on a given day or from a given batch of crude oil as a "block."

Suppose, as before, that we have v varieties and, further, that we have b blocks with one plot of each variety in each block. We modify our model in the following way: let y_{ij} denote the yield of the ith variety of wheat in the jth block; then

$$y_{ij} = \mu + \tau_i + \gamma_j + e_{ij}.$$

Here γ_j represents the effect of the jth block; if it is a fertile block, each plot in the block receives the same bonus increase in yield; similarly, in a bad block, each variety receives the same penalty. The error variance σ^2 differs from the variance considered earlier. Then we considered the difference in fertility between two plots chosen at random from the entire field; now we are dealing with the variability between plots in the same block, the intrablock variance, which is presumably smaller.

There are now $1 + v + b$ parameters and the rank of $\mathbf{X}$ is only $v + b - 1$; we impose two side conditions, $\sum_i \tau_i = 0$ and $\sum_j \gamma_j = 0$. The least squares estimate of τ_i is $y_{i\cdot} - \bar{y}$ and the corresponding estimate of γ_j is $y_{\cdot j} - \bar{y}$ where $y_{\cdot j}$ is the average yield of the jth block and $\bar{y}$ is again the average for all bv plots. We use B_j to denote the sum of the observations in the jth block. The regression procedure now divides $\mathbf{Y'Y}$ into four components:

$$\mathbf{Y'Y} = G^2/n + S_t + S_b + S_e,$$

where $G = n\bar{y}$; S_t, the sum of squares for treatments, is given by $\sum T_i^2/b - G^2/n$ as before; S_b is similarly obtained as $\sum B_j^2/v - G^2/n$; S_e is the remainder. The estimate of the intrablock variance is $s^2 = S_e/(n - v - b + 1)$, and the usual t-tests and least significant differences for contrasts in the varieties may be derived.

In this design, in which each block contains each variety exactly once, the blocks are said to be orthogonal to the varieties. We may take from the vector space of estimable functions two orthogonal subspaces, one generated by the contrasts in the τ_i and the other by the γ_j; they are orthogonal in the sense that estimating contrasts in different subspaces are statistically independent. From a more practical standpoint, our estimate of $\psi = \mathbf{c}'\boldsymbol{\tau}$ is $\sum c_i y_{i\cdot}$, which is the estimate that we obtained without blocking. The blocking has been manifested only in the smaller error variance.

The experiment that we have just described is the randomized complete block experiment. It has been reported by some research stations to be far the most widely used experimental design, and rightly so.

We turn now to situations in which we are unable to use complete blocks, and in doing so we open the door to geometers, number theorists, and combinatorists. We shall first touch upon the use of Latin squares and then go on to consider incomplete block designs.

5. LATIN SQUARES

Let us for a moment take our chemical engineer away from his pilot plant to supervise a road test on gasolines. Suppose that we wish to compare the mileage of four gasolines. The test procedure is to drive a car over a course with a test gasoline and record the amount of gasoline used. It takes a day to complete a test run. We can take four cars and, over a period of four successive days, run all four cars with all four gasolines. The gasolines are the varieties; the cars are blocks. There is also the possibility of day-to-day variation in the weather conditions. Therefore, it would be prudent to arrange, if possible, that each gasoline be tested each day. This can be done by the use of a Latin square, which is a square array of side p in which p letters are written p times each in such a way that each letter appears exactly once in each row and exactly once in each column.

Three Latin squares of side four are given in Table I. In the first square, we can let the rows denote cars, the columns days, and the letters gasolines. Then our experimental design calls for gasoline B to be used in the second car on Monday, the first on Tuesday, the fourth on Wednesday, and third on Thursday.

TABLE I

A	*B*	*C*	*D*		*A*	*B*	*C*	*D*		*A*	*B*	*C*	*D*
B	*A*	*D*	*C*		*C*	*D*	*A*	*B*		*D*	*C*	*B*	*A*
C	*D*	*A*	*B*		*D*	*C*	*B*	*A*		*B*	*A*	*D*	*C*
D	*C*	*B*	*A*		*B*	*A*	*D*	*C*		*C*	*D*	*A*	*B*

We modify our model and, letting y_{ijk} denote the observation, if any, on the ith gasoline in the jth car on the kth working day, we write

$$y_{ijk} = \mu + g_i + c_j + w_k + e_{ijk},$$

with the usual assumptions about the error terms e_{ijk}. In this model, g_i, c_j, w_k denote the contributions of the gasoline, car, and day, and $\sum g_i = \sum c_j = \sum w_k = 0$.

To compare two gasolines, we take the differences between the averages of the four observations on each of them, and, since each letter of our square appears once in each row and each column, the car effects and day effects cancel out. Designs of this kind are often called designs for the two-way elimination of heterogeneity. The sums of squares for gasolines, cars, and days each have $(p - 1)$ d.f. They are, respectively $p^{-1} \sum T_{i..}^2 - C$, $p^{-1} \sum T_{.j.}^2 - C$ and $\sum T_{..k}^2 - C$ where $T_{i..}$, $T_{.j.}$, $T_{..k}$ are the gasoline, car and day totals and $C = G^2/p^2$. We obtain S_e by subtracting all three of these sums of squares from $\sum_i \sum_j \sum_k y_{ijk}^2 - C$; it has $(p - 1)(p - 2)$ degrees of freedom.

For this experiment our engineer will need four drivers. We can also arrange to balance out the driver differences by using the first two squares of the three squares. If we were to use Greek letters in the second square, and superimpose it upon the first square, each Latin letter would appear exactly once with a Greek letter. Thus the driver differences, if any, would cancel out. The two squares are said to be orthogonal, and this design is called a Graeco-Latin square design.

The third square is orthogonal to the other two, and could be incorporated into the design to balance out another source of variability, such as four different courses or four different speeds. The three squares form a complete set of mutually orthogonal Latin squares (MOLS). A complete set of MOLS of side p contains $p - 1$ squares. Bose (1938) has given a method of obtaining complete sets whenever p is a prime or a power of a prime, i.e., when a Galois field with p elements exists. At the other end of the scale, in the case of $p = 6$, there is not even a pair of MOLS.

6. INCOMPLETE BLOCK DESIGNS

We consider now a design for v varieties in b blocks; the ith variety appears in r_i plots, and there are k_j plots in the jth block with n plots altogether; $\mathbf{R}$ is the diagonal matrix with the ith diagonal element r_i, and $\mathbf{K}$ is a diagonal matrix with k_j as the jth diagonal element. The incidence matrix of the design, $\mathbf{N} = (n_{ij})$, has v rows and b columns; n_{ij} is equal to the number of times that the ith variety appears in the jth block. It follows that $\mathbf{N1} = \mathbf{R1}$, and $\mathbf{N'1} = \mathbf{K1}$. For the moment, we allow the possibility that a variety may appear more than once in a block, and modify the model of the fourth section to let y_{ijm} denote the mth plot with the ith variety in the jth block; then

$$y_{ijm} = \mu + \tau_i + \gamma_j + e_{ijm},$$

with the usual distributional requirements that e_{ijm} are independent $N(0, \sigma^2)$ variables.

The normal equations may be written in the following way:

$$\begin{aligned} G &= n\hat{\mu} + \mathbf{1'R}\hat{\boldsymbol{\tau}} + \mathbf{1'K}\hat{\boldsymbol{\gamma}}, \\ \mathbf{T} &= \mathbf{R'1}\hat{\mu} + \mathbf{R}\hat{\boldsymbol{\tau}} + \mathbf{N}\hat{\boldsymbol{\gamma}}, \\ \mathbf{B} &= \mathbf{K'1}\hat{\mu} + \mathbf{N'}\hat{\boldsymbol{\tau}} + \mathbf{K}\hat{\boldsymbol{\gamma}}, \end{aligned}$$

where $\mathbf{T}$, $\mathbf{B}$ are vectors of variety totals and block totals, and $\boldsymbol{\tau}$, $\boldsymbol{\gamma}$ are vectors of variety and block effects. Eliminating $\hat{\mu}$ and $\hat{\boldsymbol{\gamma}}$, we have the intrablock equations

$$\mathbf{T} - \mathbf{NK}^{-1}\mathbf{B} = (\mathbf{R} - \mathbf{NK}^{-1}\mathbf{N'})\hat{\boldsymbol{\tau}},$$

which we write as $\mathbf{Q} = \mathbf{A}\hat{\boldsymbol{\tau}}$; $\mathbf{Q}$ is the vector of adjusted treatment totals, and $\mathbf{A}$ is called the adjusted intrablock matrix. We note that $\mathbf{A}$ is singular, since $\mathbf{A1} = \mathbf{R1} - \mathbf{NK}^{-1}\mathbf{N'1} = \mathbf{0}$; we shall assume that the rank of $\mathbf{A}$ is $v - 1$, in which case the design is said to be a connected design. The variance-covariance matrix of the adjusted treatment totals is given by cov $\mathbf{Q} = \mathbf{A}\sigma^2$.

To obtain a solution vector to the equations subject to $\mathbf{1'}\boldsymbol{\tau} = 0$, we write $\boldsymbol{\Omega}^{-1} = \mathbf{A} + a\mathbf{J}$, where $\mathbf{J} = \mathbf{11'}$ and a is a convenient nonzero scalar. Then $\boldsymbol{\Omega}$ is a generalized inverse of $\mathbf{A}$ and $\hat{\boldsymbol{\tau}} = \boldsymbol{\Omega}\mathbf{Q}$

is a solution vector. Furthermore, if $\psi = \mathbf{c}'\boldsymbol{\tau}$ is any contrast $V(\hat{\psi}) = \mathbf{c}'\boldsymbol{\Omega}\mathbf{A}\boldsymbol{\Omega}\mathbf{c}\sigma^2 = \mathbf{c}'\boldsymbol{\Omega}\mathbf{c}\sigma^2$.

Henceforth, we shall restrict our discussion to binary, equireplicate, proper designs: binary means $n_{ij} = 0$ or 1, i.e., no variety appears more than once in any block; equireplicate and proper mean that $r_i = r$ and $k_j = k$ for all i and all j, respectively. Then $\mathbf{NN}'$ has r along the diagonal; it is customary to denote the off-diagonal elements by λ_{hi}; λ_{hi} is equal to the number of blocks in which both variety h and variety i appear.

What do we seek in our choice of an incomplete block design? If we assume that no variety is more interesting to us than the others (which might not be true if one were a standard or control), one reasonable yardstick for a design is the variance of simple comparisons, $\hat{\tau}_h - \hat{\tau}_i$.

If $V(\hat{\tau}_h - \hat{\tau}_i)$ is the same for all pairs h, i, we say that the design is balanced. A necessary and sufficient condition for this is that the off-diagonal elements of $\mathbf{A}$ shall be equal. Since $\mathbf{A} = r\mathbf{I} - \mathbf{NN}'/k$, this implies that $\lambda_{hi} = \lambda$ for all pairs, and that we have what is called a balanced incomplete block design (BIBD).

7. BALANCED INCOMPLETE BLOCK DESIGNS

A balanced incomplete block design (BIBD) is a binary proper equireplicate design in which every pair of varieties appear together in exactly λ blocks. They were first introduced by Yates (1936); the first person to make a thorough investigation of their existence and construction was R. C. Bose (1939). The five parameters v, b, r, k, λ satisfy two equalities, the first of which, $rv = bk$, has already been mentioned. The second equality is $\lambda = r(k - 1)/(v - 1)$: this follows from the fact that any variety appears in r blocks, and the remaining $r(k - 1)$ plots in those blocks must contain the other varieties λ times each. Some authors use the expression $B(v, k, \lambda)$ designs to describe these designs.

Yates introduced two series of designs, called the orthogonal series, which are constructed from sets of mutually orthogonal Latin squares. The design for sixteen varieties with $b = 20$, $k = 4$

TABLE II

1 2 3 4	1 5 9 13	1 6 11 16	1 7 12 14	1 8 10 15
5 6 7 8	2 6 10 14	2 5 12 15	2 8 11 13	2 7 9 16
9 10 11 12	3 7 11 15	3 8 9 14	3 5 10 16	3 6 12 13
13 14 15 16	4 8 12 16	4 7 10 13	4 6 9 15	4 5 11 14

is given in Table II. The blocks are given in sets of four (the blocks are the rows), and each set contains each variety once; such a design is said to be resolvable (into distinct replicates).

To obtain the second set of blocks, we transpose the array formed by the first four blocks. To obtain the third array, we superimpose the first of the Latin squares of side four given in Table I upon the array. Then the varieties corresponding to A constitute one block, those corresponding to B form the next block, and so on. The last two sets of blocks are obtained in the same way from the other two Latin squares.

We now take five more varieties, 17, 18, 19, 20, 21, and add a plot with the $(16 + i)$th variety to each block in the ith set; then the addition of one more block containing the five new varieties gives us a design with $v = b = 21$, $r = k = 5$, $\lambda = 1$. In general, these series give $B(s^2, s, 1)$ and $B(s^2 + s + 1, s + 1, 1)$ designs whenever a complete set of MOLS exists.

The two conditions, coupled with the requirement that λ be an integer, are necessary but not sufficient for the existence of a design. A design with $v = b$ is said to be symmetric, and the following non-existence theorem for symmetric designs was proved independently by several authors.

THEOREM: *A necessary condition for the existence of a symmetric* BIBD *with v even is that $k - \lambda$ be a square.*

To prove this, we note that the value of the determinant of $\mathbf{NN}'$ is $(r - \lambda)^{v-1}rk$. If the design is symmetric, this quantity becomes $k^2(k - \lambda)^{v-1}$; since $\mathbf{N}$ is now a square matrix, this must equal the square of the determinant of $\mathbf{N}$. Hence, $(k - \lambda)^{v-1}$ must be a square. This proves, for example, that there is no design with $v = b = 22$, $r = k = 7$, $\lambda = 2$.

Another example of nonexistence is the case of $v = 36$, $k = 6$, $\lambda = 1$. It can be shown that such a design would have to belong to Yates' orthogonal series, but there is no set of MOLS of side 6.

The construction of designs is an active area of research. In his paper, Bose (1939) considered all sets of parameters satisfying the necessary conditions and having $r \leqslant 10$, $k \leqslant 10$. He was able to solve all but twelve cases, either by finding the design or by establishing that it did not exist. Subsequent workers have solved ten of the cases, including the symmetric design for 22 varieties mentioned above. Two cases remain unsolved:

(1) $v = 46 \quad b = 69 \quad r = 9 \quad k = 6 \quad \lambda = 1;$
(2) $v = 51 \quad b = 85 \quad r = 10 \quad k = 6 \quad \lambda = 1.$

We do not know whether designs with these parameters exist.

Attention has also turned to the existence of nonisomorphic solutions for various sets of parameters. For example, Nandi (1946) showed that there are exactly four nonisomorphic $B(8, 4, 3)$ designs. Two designs are said to be isomorphic if we can obtain one from the other by relabeling varieties and reordering the blocks. One reason for the interest in nonisomorphic solutions lies in the resistance of some designs to the loss of a variety. Hedayat and John (1974) considered the effect upon a BIBD when one of the varieties has to be discarded; it might, for example, be the only variety that is susceptible to a spray that was accidentally used on the field and its plots are lost from the experiment. Hedayat and John investigated conditions under which the remaining plots form a balanced design for $v - 1$ varieties, and found that it was necessary and sufficient that the blocks of size k and the blocks of size $k - 1$ each constitute a BIBD. Only one of Nandi's four solutions has this property for the loss of any one of its varieties.

8. PARTIALLY BALANCED INCOMPLETE BLOCK DESIGNS

The requirement that λ must be an integer restricts severely the number of balanced designs, so that, although balance is a desirable property, we are forced as a practical matter to consider other designs in planning experiments. We showed earlier that, for any contrast $\psi = \mathbf{c}'\boldsymbol{\tau}$, $V(\hat{\psi}) = \mathbf{c}'\boldsymbol{\Omega}\mathbf{c}\sigma^2$, and this focuses our attention upon the pattern of the matrix $\boldsymbol{\Omega}$. Pearce (1963) suggested in a survey paper on the choice of designs that we might characterize incomplete block designs by the pattern of $\boldsymbol{\Omega}^{-1}$. We do not in the general case know what the pattern of $\boldsymbol{\Omega}$ is going to be, but, since it differs only by the addition of a diagonal matrix, the pattern of $\boldsymbol{\Omega}^{-1}$ is the same as that of $\mathbf{NN}'$. We shall now show that for a large class of designs called partially balanced designs the structure of $\boldsymbol{\Omega}$ is, in an important sense, the same as the structure of $\mathbf{NN}'$.

Partially balanced designs were introduced by Bose and Nair (1939). Since then their construction, existence, and general properties have been subjects of active research by Bose, his students, and others. A valuable survey of the mathematical properties was written by Bose (1963) in the memorial volume for Professor Mahalanobis. They are a special class of designs in which some varieties appear together in λ_1 blocks, others in λ_2 blocks, and so on.

We begin by defining a partially balanced association scheme for v varieties with m association classes. This is an arrangement whereby

(i) any two varieties are first, second, . . ., or mth associates (this relation is symmetric so that if α is an ith associate of β, then β is an ith associate of α);

(ii) each variety has exactly n_i ith associates;

(iii) if α and β are ith associates, the number of varieties that are both jth associates of α and kth associates of β is denoted by p^i_{jk} and is independent of the pair of ith associates chosen.

It follows that $p^i_{jk} = p^i_{kj}$ and that $\sum_{i=1}^{m} n_i = v - 1$. It is convenient to call any variety its own zeroth associate and to define $p^0_{ii} = n_i$, and $p^0_{ij} = 0$, if $i \neq j$.

A partially balanced design with this scheme is a proper, balanced, equireplicate design in which each pair of ith associates appears together in exactly λ_i blocks. As an example, consider the following scheme for eight varieties. We number the varieties 0 through 7 and divide them into four groups of two each: 04, 15, 26, 37. Two varieties are first associates if they are in the same group; otherwise, they are second associates. This is called a group divisible scheme. The following design with $b = 8$ and $k = 3$ has $\lambda_1 = 0$ and $\lambda_2 = 1$:

016, 127, 230, 341, 452, 563, 674, 705.

We note, in passing, that this is a cyclic design. We start with an initial block 016; the $(j + 1)$th block is obtained by adding j to each member of the initial block and reducing modulo 8.

The association matrices of a scheme are a set of $m + 1$ matrices, $\mathbf{B}_i$, each having v rows and v columns. The entry in row α and column β of $\mathbf{B}_i$ is unity if α and β are ith associates and zero otherwise. The matrices are symmetric with the row sums of $\mathbf{B}_i$ each equal to n_i; $\mathbf{B}_0$ is the identity matrix $\mathbf{I}_v$. The association matrices sum to $\sum_{i=0}^{m} \mathbf{B}_i = \mathbf{J} = \mathbf{11}'$. They are linearly independent in that $\sum_i c_i\mathbf{B}_i = \mathbf{0}$ if, and only if, all the c_i are zero. The linear combinations of these matrices form an $(m + 1)$-dimensional vector space of which the association matrices themselves constitute a basis.

Products of association matrices are also members of the vector space, since $\mathbf{B}_j\mathbf{B}_k = \sum_i p_{jk}^i\mathbf{B}_i$, and so the linear combinations of these matrices form a ring with unit element $\mathbf{I}$.

If we set $\lambda_0 = r$, we may write $\mathbf{NN}'$ as $\sum_{i=0}^{m} \lambda_i\mathbf{B}_i$, and $k\mathbf{A}$ as $rk\mathbf{B}_0 - \sum_{i=0}^{m} \lambda_i\mathbf{B}_i$. We say that $\mathbf{\Omega}$ has the same pattern as $\mathbf{NN}'$ if $\mathbf{\Omega}$ is also a linear combination of the association matrices. This is clearly the case, since

$$\mathbf{\Omega}^{-1} = (\mathbf{A} + a\mathbf{J}) = (r + a)\mathbf{B}_0 + \sum_{i=0}^{m} (a - \lambda_i/k)\mathbf{B}_i.$$

A consequence of this result, namely that $\mathbf{\Omega} = \sum_{i=0}^{m} \mu_i\mathbf{B}_i$, is that if α and β are ith associates, $V(\hat{\tau}_\alpha - \hat{\tau}_\beta) = 2(\mu_0 - \mu_i)\sigma^2$; i.e., all comparisons between ith associates have the same variance, and so $V(\hat{\tau}_\alpha - \hat{\tau}_\beta)$ can take only m different values. For the balanced incomplete block designs, $V(\hat{\tau}_\alpha - \hat{\tau}_\beta)$ takes only one value; for the partially balanced design with two associate classes, there are two possible values, and so on.

9. FACTORIAL EXPERIMENTS

Suppose that our chemical engineer wishes to explore the effect of varying temperature and pressure in his plant upon some aspect of the product, which we shall call the yield. Suppose further that he has chosen t different temperatures at which to run his plant and p different pressures. One possible experimental design is to make r runs at each of the tp possible combinations of temperature and pressure. Just as in section four every variety appeared once in each block, each temperature now appears r times with each pressure. The balance, or orthogonality, between what we now call the two factors is maintained. We could take a model very similar to the randomized block model and represent the yield on the kth of the runs made at the ith temperature and the jth pressure by

$$y_{ijk} = \mu + \tau_i + \pi_j + e_{ijk},$$

where τ_i now represents the effect of the ith level of the temperature factor and π_j the jth level of the pressure.

The term τ_i is sometimes called the main effect of the ith level of temperature. If we denote the average yield at that level by $y_{i.}$ and $G/(rpt)$ by $\bar{y}$, we may think of τ_i as the expected value of $(y_{i.} - \bar{y})$; similarly, $\pi_j = E(y_{.j} - \bar{y})$. The side conditions $\sum_i \tau_i = 0$ and $\sum_j \pi_j = 0$ are then obvious.

The sums of squares for temperature and pressure, by which we really mean the sums of squares due to differences between temperatures and differences between pressures, are given by

$$S_t = \sum_i T_{i.}^2/(rp) - C \qquad \text{and} \qquad S_p = \sum_j T_{.j}^2/(rt) - C,$$

respectively, where $T_{i.}$ is the sum of the yields at the ith level of temperature, $T_{.j}$ is the sum of the yields at the jth level of pressure, G is the grand total for all rpt runs and $C = G^2/(rpt)$. The sums of squares have $t - 1$ and $p - 1$ degrees of freedom, respectively.

This simple additive model has a major drawback. It implies that at every level of pressure the expected increase in yield when we change temperature from level i to level i' is the same, namely $\tau_{i'} - \tau_i$; it also implies a similar constancy when we change from one level of pressure to another at every temperature. If we were to consider our experiment as a one-way layout in which we made r runs at each of $v = pt$ sets of operating conditions, we should have an error sum of squares S_e with $pt(r - 1)$ degrees of freedom and what we might call a sum of squares for "operating conditions" with $pt - 1$ degrees of freedom, which we shall denote by S_v. We have $S_v = \sum_i \sum_j T_{ij}^2/r - C$, where T_{ij} is the sum of the yields at the ith temperature and the jth pressure, i.e., in the (ij)th cell.

How do we explain the difference between the two error terms S_e and $S'_e = \sum\sum\sum (y_{ijk} - \bar{y})^2 - S_t - S_p$ in terms of the regression models? We add pt new terms to our additive model and write

$$y_{ijk} = \mu + \tau_i + \pi_j + (\tau\pi)_{ij} + e_{ijk}, \qquad i = 1, \ldots, t; j = 1, \ldots, p;$$

$(\tau\pi)_{ij}$ is called the interaction of the ith level of temperature and the jth level of pressure. It represents the deviation of the yield at that set of operating conditions from that given by the simple additive model. There are two sets of appropriate side conditions for the new parameters: $\sum_j (\tau\pi)_{ij} = 0$ for each i and $\sum_i (\tau\pi)_{ij} = 0$ for each j. One of these side conditions is redundant, since both sets imply $\sum_i \sum_j (\tau\pi)_{ij} = 0$, and so we have only added $t + p - 1$ new conditions. Our model now contains $1 + t + p + pt$ parameters, and the rank of $\mathbf{X}'\mathbf{X}$ is pt.

To test the adequacy of the simple additive model, we test the hypothesis that the set of interaction parameters are all zero. The corresponding sum of squares is $S_v - S_t - S_p$ and the F statistic has $(pt - 1) - (t - 1) - (p - 1) = (t - 1)(p - 1)$ degrees of freedom in the numerator.

More generally, if we have two factors A and B with a and b levels, respectively, and r observations at each combination of the two factors, we may summarize the data in the following analysis of variance table:

Source	d.f.	Sum of Squares
A	$a - 1$	$S_A = \sum_i T_{i.}^2/(br) - C$
B	$b - 1$	$S_B = \sum_j T_{.j}^2/(ar) - C$
AB (interaction)	$(a - 1)(b - 1)$	$S_{AB} = \sum_i \sum_j T_{ij}^2/r - C - S_A - S_B$
Error (within cells)	$ab(r - 1)$	$S_e = \sum_i \sum_j \sum_k (y_{ijk} - \bar{y})^2 - S_A - S_B - S_{AB}$.

With more factors we have higher order interactions. An experiment with four factors will have in its model and in its analysis of variance table terms for six two-factor interactions (*AB*, *AC*, *AD*, *BC*, *BD*, *CD*), for four three-factor interactions (*ABC*, *ABD*, *ACD*, *BCD*), and for one four-factor interaction. For example, the *ABC* interaction has $(a-1)(b-1)(c-1)$ d.f. and the four-factor interaction has $(a-1)(b-1)(c-1)(d-1)$ d.f.

10. EXPERIMENTS WITH FACTORS AT TWO LEVELS

An experiment with n factors, each of which has only two levels, is called a 2^n factorial. These experiments were first considered systematically by Yates (1937). There are 2^n sets of experimental conditions, which are usually called treatment combinations in the jargon of the topic. It will be simplest for us to focus upon the 2^3 experiment; generalization to more than three factors will be obvious.

Suppose that our chemical engineer has three factors to consider: temperature, pressure, and flow rate (measured in gallons per hour), which we label *A*, *B*, *C*, respectively. We rescale the factors so that the levels of each become ± 1; that is, we associate with the ith factor the variable x_i which takes the value $+1$ when the factor is at its high level and -1 at the low level. The set of experimental points is now the set of vertices of a cube in three dimensional Euclidean space with coordinates ± 1.

We take as our model

$$\begin{aligned} y = \beta_0 + \beta_1 x_1 + \beta_2 x_2 + \beta_3 x_3 + \beta_{12} x_1 x_2 \\ + \beta_{13} x_1 x_3 + \beta_{23} x_2 x_3 + \beta_{123} x_1 x_2 x_3 + e; \end{aligned}$$

β_i is the main effect of the ith factor, β_{ij} is the two factor interaction, and β_{123} the three factor interaction. It is customary to denote them also by *A*, *AB*, *ABC*, and so on. We shall assume, in this section and the next, that one run is made with each of the 2^3 treatment combinations.

Yates introduced a useful notation for the treatment combinations. We denote the point at which *A* is at its high level and the other factors are at their low levels by *a*; *bc* denotes the point at which *A* is at its low level and *B*, *C* are at their high levels; the point at which every factor takes its low level is denoted by (1). We shall use these symbols to denote either the points themselves or the values of the response at the points. This gives us another mapping of the points in the sample space from the vertices of the cube to the elements of a cyclic group of order 2^3 with (1) as identity and $a^2 = b^2 = c^2 = (1)$; changing the level of temperature from one run to another corresponds to multiplying the representation of the run in the group by a.

With one run at each of the 2^n points, $\mathbf{X}$ is a Hadamard matrix, $\mathbf{X}'\mathbf{X}$ takes the simple form $2^n\mathbf{I}$, and we have $2^n\hat{\boldsymbol{\beta}} = \mathbf{X}'\mathbf{Y}$,

$$2^n\hat{\beta}_0 = \sum y, \qquad 2^n\hat{\beta}_i = \sum x_i y, \qquad 2^n\hat{\beta}_{ij} = \sum x_i x_j y, \ldots.$$

In particular, $8\hat{\beta}_1 = abc + ab + ac + a - bc - b - c - (1)$. This is a contrast in the observations and is called the A contrast; it is, as we might have expected, the difference between the sum of the responses at the high level of temperature and the sum of the responses at the low level or the total at $x_1 = +1$ minus the total at $x_1 = -1$. The AB contrast, $8\hat{\beta}_{12}$, can be looked at in two ways. As $\sum x_1 x_2 y$, we see it as the difference between the totals at $x_1 x_2 = +1$ and at $x_1 x_2 = -1$. We may also write it as $\{abc + ab - bc - b\} - \{ac + a - c - (1)\}$; the contrast in the first brackets gives us the effect of changing temperature at high pressure, while the contrast in the second brackets gives the effect at low pressure; their difference is the interaction. Symbolically, we may write the two contrasts as $(a - 1)(b + 1)(c + 1)$ and $(a - 1)(b - 1)(c + 1)$, respectively.

The estimating contrasts are mutually orthogonal. With only as many points as we have parameters, there is not, as our model stands, any error term. We have two options: we may use a prior estimate of error, which is not unreasonable when we have experience with similar experiments on the plant, or we may elect to suppress the higher order interactions from our model. We usually assume as a practical matter that, in the absence of prior indications to the contrary, we can omit interactions with three or more factors. That procedure is not much help with the 2^3 experiment, but with five factors it gives us an error term with sixteen degrees of freedom.

11. CONFOUNDING

Suppose that we are obliged to run our 2^3 experiment in two blocks of four runs each. How should we assign the treatment combinations to the blocks? It would not be a good idea to put all the runs at high temperature into one block and the runs at low temperature into another, because then the contrast for the main effect of temperature would be identical with the difference between the block totals. If this contrast were large, we should not know whether that was due to a temperature effect or to block differences. In that assignment, the temperature effect is said to be confounded with blocks.

A better idea is to allocate the treatment combinations so as to do the least damage. In this experiment we could choose to confound *ABC*, which is the highest order interaction. This calls for putting (1), *ab*, *ac*, *bc* into one block, and *a*, *b*, *c*, *abc* into the other block, i.e., we put into one block the runs with $x_1x_2x_3 = +1$, and into the other the runs with $x_1x_2x_3 = -1$. By doing this, we have lost our ability to estimate *ABC* but, since the other estimating contrasts are orthogonal to the *ABC* contrast, we may estimate the other effects as before, unaffected by the blocking. In estimating β_1, for example, we take $\{abc + ab + ac + a\} - \{bc + b + c + (1)\}$; each block contains two runs at high temperature and two at low temperature, and the block differences cancel out.

This gives us a general rule for confounding in a 2^n design with two blocks of size 2^{n-1}. We confound the highest order interaction by putting into one block the runs with $x_1x_2 \cdots x_n = +1$ and into the other the runs with $x_1x_2 \cdots x_n = -1$.

What if we wish to divide our factorial into four blocks of size 2^{n-2} (or more generally s^k blocks of size 2^{n-k})? There are three degrees of freedom for block differences. Can we choose any three interaction contrasts to equate to interblock differences? We shall see that this cannot be done.

Suppose that we wish to divide a 2^4 factorial into four blocks of four runs each, and that we elect to confound *ABC* and *ABCD*. This defines our blocks as the subsets I$\{x_1x_2x_3x_4 = 1, x_1x_2x_3 = -1\}$, II$\{x_1x_2x_3x_4 = +1, x_1x_2x_3 = +1\}$, III$\{x_1x_2x_3x_4 = -1, x_1x_2x_3 = +1\}$, IV$\{x_1x_2x_3x_4 = -1, x_1x_2x_3 = -1\}$.

However, if $x_1x_2x_3x_4 = 1$ and $x_1x_2x_3 = -1$, $x_1^2x_2^2x_3^2x_4 = -1$ which implies, since $x_i^2 \equiv 1$, $x_4 = -1$; thus our blocks contain the subsets I$\{x_4 = -1\}$, II$\{x_4 = +1\}$, III$\{x_4 = -1\}$, IV$\{x_4 = +1\}$, and we have also confounded the main effect D.

We now introduce the idea of the group of effects. The term effect is used to denote any main effect or interaction. With the addition of an identity, I, the set of effects also forms an Abelian group of order 2^n with $A^2 = B^2 = C^2 = \cdots = I$ (corresponding to $x_i^2 \equiv 1$). If P and Q are any two effects, their product in the group is called their generalized interaction. Our example illustrates that when $ABCD$ and ABC are confounded, so is their product $A^2B^2C^2D = D$. In general, if P and Q are confounded, so is their product, and we now have the following result: the set of effects that are confounded, together with the identity, form a subgroup.

Ideally, we try, in setting up a confounding pattern, to avoid confounding main effects and two-factor interactions, but sometimes our hands are tied. If, for the moment, we consider the members of the subgroup, except for I, as words, and call the number of letters in an effect the length of that word, we can denote the length of the ith word by w_i. It can then be shown that, if m out of the n letters occur in the subgroup and if the subgroup has 2^k elements, each of the m letters appears in exactly 2^{k-1} words, so that $\sum w_i = m2^{k-1}$.

Returning to our example with $n = 4$, $k = 2$, we should, if we want to make our words as long as possible, include all four letters in the subgroup. Even then we have $\sum w_i = 8$. There are three words, and there is no way in which we can have each word contain at least three letters; we must confound at least one main effect or two-factor interaction. We may, for example, confound ABC, ABD, and CD (see Table III), but if we elected to confound the four-factor interaction and a three-factor interaction, we should be forced to confound a main effect.

We can, however, divide the 2^5 design into four blocks of size eight, confounding only higher order interactions, by choosing a four-factor interaction and a pair of three-factor interactions. If we choose to confound $ABCD$, ABE, CDE, the block defined by $x_1x_2x_3x_4 = 1$ and $x_1x_2x_5 = -1$ consists of (1), *ade*, *bde*, *ab*, *cd*, *ace*, *bce*, *abcd*, which is a subgroup of the group of treatment

TABLE III

1	*ab*	*bcd*	*acd*
cd	*abcd*	*b*	*a*
abc	*c*	*ad*	*bd*
abd	*d*	*ac*	*bc*

combinations. The block which contains the basepoint (1) is called the principal block. When a 2^n design is confounded in 2^k blocks of size 2^{n-k}, the principal block is a subgroup and the other blocks are its cosets.

The elimination of two-way heterogeneity can be handled by a system of double confounding. The treatment combinations of the 2^4 design are arranged in a 4×4 array in Table III. Considering the rows as blocks, we have the division into four blocks confounding *ABC*, *ABD*, and *CD*. We also have confounded *ACD*, *BCD*, and *AB* between columns, which leaves the main effects, the remaining two-factor interactions, and the four-factor interaction unconfounded.

12. PARTIAL CONFOUNDING AND FACTORIAL STRUCTURE

We have seen that if we are only able to make one run at each treatment combination, some must be sacrificed by being confounded with block differences. If we are able to repeat the experiment several times, we can choose different subsets of effects to confound each time, so that every treatment is estimable from at least some of the replicates. If we were able to repeat the 2^3 experiment seven times, each time in two blocks of four runs each, we could confound each of the seven effects in a single replicate, and estimate it from the remaining six. Each effect would then be estimated from 48 points with $V(\hat{\beta}) = \sigma^2/48$.

The actual design is given in Table IV: *A* is confounded in the first replicate, *B* in the second, and so on. *A* is estimated by averaging the estimates from the last six replicates, and *B* is estimated by averaging over the first replicate and the last five. Since the *A* and *B* contrasts are independent in each of the last five replicates, $\hat{\beta}_1$

TABLE IV

(1) b c bc	(1) a c ac	(1) a b ab	(1) ab c abc	(1) ac b abc	(1) bc a abc	(1) ab ac bc
A	B	C	AB	AC	BC	ABC
a ab ac abc	b ab bc abc	c ac bc abc	a b ac bc	a c ab bc	b c ab ac	a b c abc

and $\hat{\beta}_2$ are also independent. Indeed, the estimates of all seven effects are mutually independent.

The design that we have used is a balanced incomplete block design with $v = 8$, $b = 14$, $r = 7$, $k = 4$, $\lambda = 3$. We mentioned earlier that there are four nonisomorphic solutions to the problem of finding a BIBD with those parameters. The design that we have chosen has an obvious structure, but we may ask whether any of the other three BIB designs with these parameters would do as well. We ask that question in two senses: first, would the $\hat{\beta}_i$ still have $V(\hat{\beta}_i) = \sigma^2/48$? second, would the estimates still be mutually orthogonal? We can now use some of the results of the previous section to show that the designs are equivalent in both senses.

We first consider the design as a BIBD with the eight treatment combinations as the varieties. For any of the BIB designs with these parameters, $\mathbf{NN'}$ has $r = 7$ along the main diagonal and $\lambda = 3$ elsewhere, so that $\mathbf{A} = 3(8\mathbf{I} - \mathbf{J})/4$. If we now write $\mathbf{\Omega}^{-1} = \mathbf{A} + 3\mathbf{J}/4 = 6\mathbf{I}$, we have $\mathbf{\Omega} = \mathbf{I}/6$.

We now recall that $8\boldsymbol{\beta} = \mathbf{M}\boldsymbol{\tau}$, where $\mathbf{M}$ is the Hadamard matrix which played the part of the $\mathbf{X}$ matrix in the previous section. It follows that $\hat{\boldsymbol{\beta}} = \mathbf{M}\hat{\boldsymbol{\tau}}/8$, whence $\operatorname{cov} \hat{\boldsymbol{\beta}} = \mathbf{M\Omega M'}\sigma^2/64$; but $\mathbf{MM'} = 8\mathbf{I}$, so that $\operatorname{cov} \hat{\boldsymbol{\beta}} = \mathbf{I}\sigma^2/48$. This establishes that for each effect $V(\hat{\beta}) = \sigma^2/48$; also, since the covariance matrix is diagonal, the estimates are independent.

The BIB design for sixteen varieties presented in Section 7 may be used for a 2^4 factorial. If we associate the treatment combinations with the varieties by superimposing the array of Table III upon the first set of four blocks in Table II, so that $1 \to (1)$, $2 \to ab$, $3 \to bcd$, etc., the first replicate has ABC, ABD, and CD confounded with blocks. In the subsequent replicates the triples of confounded effects are

$$ACD, BCD, \text{and } AB; \quad AD, BC, ABCD; \quad A, C, AC; \quad B, D, BD.$$

For this design, each effect will be estimated from four of the replicates with $V(\hat{\beta}) = \sigma^2/64$, and again the estimates of the effects will be mutually independent.

An interesting facet of both these examples is that, even though we had incomplete block designs and the complications of non-orthogonality between treatments and blocks, nevertheless, the linear combinations that gave the estimates of $\boldsymbol{\beta}$ were independent. Does this property hold for other incomplete block designs that are not balanced? The answer to this question can be seen to be affirmative when we consider what happens if we omit the last replicate from the design in Table IV. We again estimate the effects from those replications in which they are not confounded, and the argument that we used above indicates their mutual independence. This new design happens to be a partially balanced design with two associate classes of the group divisible type.

J. A. John and Smith (1972) have defined an incomplete block design as having factorial structure for a factorial experiment if the sums of squares for A, B, AB, etc., are mutually independent quadratic forms. In the case of the 2^n design, this is equivalent to having the estimates $\hat{\boldsymbol{\beta}}$ mutually orthogonal, which, in turn, is equivalent to having $\mathbf{M\Omega M}'$ be a diagonal matrix.

In our BIBD example, the result followed directly, once it was established that $\boldsymbol{\Omega}$ was diagonal. However, for the group-divisible design, or any other unbalanced design, $\boldsymbol{\Omega}$ is not diagonal. What is needed is that the rows of $\mathbf{M}$ shall be latent vectors of $\boldsymbol{\Omega}$. We have noted in the section on partially balanced designs a particularly strong connection between the structures of $\boldsymbol{\Omega}$ and $\mathbf{NN}'$. We can show that for the general binary proper equireplicate design the latent vectors of $\mathbf{NN}'$ are also latent vectors of $\boldsymbol{\Omega}$, and we have the following general result, which again enables us to classify the designs by examination of $\mathbf{NN}'$. A necessary and sufficient condition that an incomplete block design for 2^n varieties shall have factorial structure for the 2^n factorial experiment is that the vectors $\mathbf{c}$ that define the contrasts for the effects ($\hat{\beta} = \mathbf{c}'\hat{\boldsymbol{\tau}}$), together with **1**, shall be a set of mutually orthogonal latent vectors of $\mathbf{NN}'$.

At last we have an answer to the question that we asked at the beginning of this essay. We want a design with $b = v = 8$, $k = 3$ which has factorial structure for the 2^3 design. We mentioned it in passing earlier. If we take that group divisible design for $v = b = 8$, $r = k = 3$ with $\lambda_1 = 0$ and $\lambda_2 = 1$, and associate the varieties with the treatment combinations by mapping $0 \to (1)$, $1 \to ab$, $2 \to ac$, $3 \to bc$, $4 \to abc$, $5 \to c$, $6 \to b$, $7 \to a$, we have the following design: (1) *ab b*, *ab ac a*, *ac bc* (1), *bc abc ab*, *abc c ac*, *c b bc*, *b a abc*, *a* (1) *c*, which looks, at first glance, somewhat devoid of system. If we now order the varieties in $\mathbf{NN}'$ by groups, i.e., 0 4 1 5 2 6 3 7, we have a matrix which is cyclic in 2×2 submatrices. So are $\boldsymbol{\Omega}^{-1}$ and $\boldsymbol{\Omega}$, and factorial structure is then easily demonstrated. It will be noted that we have plenty of latitude in the assignment of treatment combinations to varieties. All that we need is that the image of 3 should be the product of the images of 1 and 2, and that the images of 5, 6, 7 should be the products of the images of 1, 2, and 3, respectively, with the image of 4.

13. CONCLUSION

We have touched upon only a part of the subject of experimental design. We have said nothing of the geneticist working with components of variance and heritabilities, or of mixed, fixed, and random models. No mention has been made of the work of G. E. P. Box and his colleagues in fitting response surfaces for chemical processes.

For further information about current combinatorial research in block designs, the reader is referred to the article by Marshall Hall, Jr., in *MAA Studies in Mathematics, Volume 17, Studies in Combinatorics* (Gian-Carlo Rota, editor).

The areas of application, each of which has its own special facets but still employs the same basic principles, increase daily. The mathematician who has hitherto thought of statistical journals only in terms of the *Annals of Statistics*, the *Journal of the Royal*

Statistical Society, and the *Journal of the American Statistical Association* can profitably broaden his range by looking at such journals as *Biometrika*, *Technometrics*, and *Biometrics*. They will give him some new insights into experimental design as well as other topics. I hope that he will be pleasantly surprised, and a little excited, by the wide range of interesting problems that he will discover.

BIBLIOGRAPHY

1. R. C. Bose, "On the application of the properties of Galois fields to the problem of the construction of hyper-Graeco-Latin squares," *Sankhyā*, **3** (1938), 323–338.
2. ———, "On the construction of balanced incomplete block designs," *Ann. of Eugenics*, **9** (1939), 353–399.
3. ———, "Combinatorial properties of partially balanced designs and association schemes," *Sankhyā* Ser. A., **25** (1963), 109–136.
4. R. C. Bose and K. R. Nair, "Partially balanced incomplete block designs," *Sankhyā*, **4** (1939), 337–372.
5. D. R. Cox, *Planning of Experiments*, Wiley, New York, 1958.
6. A. Hedayat and P. W. M. John, "Resistant and susceptible BIB designs," *Ann. Statist.*, **2** (1974), 148–158.
7. J. A. John and T. M. F. Smith, "Two-factor experiments in non-orthogonal designs," *J. Roy. Statist. Soc.* Ser. B., **34** (1972), 401–410.
8. P. W. M. John, *Statistical Design and Analysis of Experiments*, Macmillan, New York, 1971.
9. O. Kempthorne, *The Design and Analysis of Experiments*, Wiley, New York, 1952.
10. R. G. Miller, *Simultaneous Statistical Inference*, McGraw-Hill, New York, 1966.
11. H. K. Nandi, "A further note on non-isomorphic solutions of balanced incomplete block designs," *Sankhyā*, **7** (1946), 313–316.
12. S. C. Pearce, "The use and classification of non-orthogonal designs," *J. Roy. Statist. Soc.* Ser. A., **126** (1963), 353–369.
13. G.-C. Rota (editor), *MAA Studies in Mathematics, Volume 17, Studies in Combinatorics*, Mathematical Association of America, 1978.
14. H. Scheffé, *The Analysis of Variance*, Wiley, New York, 1959.
15. F. Yates, "Incomplete randomized blocks," *Ann. of Eugenics*, **7** (1936), 121–140.
16. ———, *The Design and Analysis of Factorial Experiments*, Imperial Bureau of Soil Science, Harpenden, England, 1937.

A BRIEF SURVEY OF NONPARAMETRIC STATISTICS

Gottfried E. Noether

1. INTRODUCTION

Most standard statistical procedures assume either explicitly or implicitly that observations are normally distributed. If the assumption of normality is satisfied, the given procedures enjoy certain optimality properties. In general, however, if the normality assumption is not satisfied, the standard procedures may be far from optimal. To give a very simple example, the best estimate of the center of a normal distribution is the sample mean. But in nonnormal populations the sample mean can be a very bad or even useless estimate of a population characteristic such as the center of symmetry.

Another drawback of normal theory methods is the inaccuracy of the nominal probability statements associated with them. In the case of confidence intervals, the true confidence coefficient may differ considerably from the nominal confidence coefficient suggested by normal theory. Correspondingly, in the case of tests of hypotheses, the true probability of an erroneous decision may differ considerably from the probability predicted by normal theory. For these and other reasons statisticians have developed alternative methods of statistical analysis that require much less stringent assumptions for their validity than normal theory methods. These methods are commonly known as *nonparametric* methods.

In the statistical literature the term *nonparametric* is used rather indiscriminately. Originally the term nonparametric is due to Wolfowitz [15], who suggested it in order to indicate that the populations under consideration could not be specified in terms of a finite number of parameters. In a sense, the term nonparametric is unfortunate. As we shall see, nonparametric methods can be used to find confidence intervals for parameters like the median of a distribution. In this article, as in most other publications on the subject, the term nonparametric remains vague.

The field of nonparametric statistics is vast. Of necessity the present article covers only certain limited aspects. We shall use the two-sample problem to explore basic ideas such as the distribution-free nature of nonparametric procedures, consistency, efficiency and problems caused by ties among the observations. The one-sample problem, *c*-sample problem, regression and correlation are discussed much more briefly.

2. THE TWO-SAMPLE PROBLEM

The most extensively studied problem in nonparametric statistics is the two-sample problem: Given two independent random samples $X_1, \ldots, X_m$ and $Y_1, \ldots, Y_n$ from populations with cumulative distribution functions (c.d.f.) $F(z)$ and $G(z)$, respectively, we want to test the hypothesis

$$G(z) = F(z), \tag{1}$$

where $F(z)$ is unspecified. The hypothesis (1) states that the two samples have come from one and the same population.

When testing a parametric hypothesis, we expect an appropriate test to reject the hypothesis when it is false with probability arbitrarily close to 1 provided the sample size is sufficiently large. For nonparametric hypotheses like (1), the situation is usually not as simple. We shall see that there exist tests that are effective against the completely general alternative

$$G(z) \neq F(z). \tag{2}$$

But the statistician is usually able to increase the effectiveness of his test by restricting the class of alternatives. Two alternatives that have been studied extensively in connection with the two-sample problem are stochastic ordering and shift.

Stochastic ordering:

$$G(z) \leqslant F(z), \qquad \text{all } z. \tag{3}$$

If X is a random variable having distribution $F(z)$ and Y, a random variable having distribution $G(z)$, for every z,

$$P(Y \leqslant z) = G(z) \leqslant F(z) = P(X \leqslant z).$$

Alternative (3) is appropriate when the statistician suspects or hopes that X-observations tend to be smaller than Y-observations.

Shift alternatives:

$$G(z) = F(z - \Delta), \tag{4}$$

where Δ is an unknown constant. Under (4),

$$P(Y \leqslant z) = G(z) = F(z - \Delta) = P(X \leqslant z - \Delta) = P(X + \Delta \leqslant z).$$

Random variables Y and $X + \Delta$ have identical distributions. This is a special case of (3). Model (4) generalizes the normal theory setup where we are interested in the difference $\Delta = \mu_y - \mu_x$ of means of two normally distributed populations and are willing to make the assumption that the two normal distributions have the same standard deviation.

Alternative (4) is a typical case when we may want to estimate a parameter, namely Δ, using "nonparametric" techniques. We shall do so in Section 7.

3. HYPOTHESIS TESTING

Classical normal-theory tests depend on the actual sample observations. Most nonparametric tests depend on the observations only through their ranks.

Given observations $Z_1, \ldots, Z_N$, the rank $r[Z_k]$ of the observation Z_k is obtained by arranging all N observations from the smallest to the largest and assigning the rank 1 to the smallest observation, the rank 2 to the second smallest observation, and so on. This procedure assumes that no ties occur among the N observations, i.e., that no two or more observations take the same value. This condition will be satisfied with probability 1 if the observations have continuous distributions. We shall make this assumption for the time being to avoid complications, even though it is unrealistic in practice. In Section 8 we discuss how to deal with tied observations.

An "obvious" rank statistic for the two-sample problem is the difference between sample rank averages

$$\frac{1}{n}\sum_{j=1}^{n} r[Y_j] - \frac{1}{m}\sum_{i=1}^{m} r[X_i], \tag{5}$$

where ranks are computed relative to the combined set of X- and Y-observations so that here $N = m + n$. Actually, since $\sum_i r[X_i] + \sum_j r[Y_j] = N(N + 1)/2$, the test based on the sum W of Y-ranks is completely equivalent to the test based on (5). The W-test is known as the Wilcoxon two-sample test, and is perhaps the most widely used procedure in all of nonparametric statistics.

In Section 10 we shall see that the Wilcoxon test is one of a large class of tests, the class of linear rank tests. However, our exploration of basic ideas will be based primarily on the Wilcoxon test.

We start by simplifying our notation and deriving an alternative form for the Wilcoxon test, which is often more convenient. Let $R_j = r[Y_j]$, $j = 1, \ldots, n$, denote the rank of the jth Y-observation in the combined set of X- and Y-observations. Then

$$W = \sum_{j=1}^{n} R_j. \tag{6}$$

The Wilcoxon statistic (6) takes its minimum value $1 + 2 + \cdots + n$

$= n(n + 1)/2$ when all n Y-observations are smaller than all m X-observations. Interchanging the largest Y-observation with the smallest X-observation increases the value of W by one. Continuing by single steps, we see that

$$W = n(n + 1)/2 + U,$$

where

$$U = \sum_{i=1}^{m} \sum_{j=1}^{n} V_{ij} \tag{7}$$

and

$$V_{ij} = \begin{cases} 1 \\ 0 \end{cases} \text{if } X_i \begin{cases} < \\ > \end{cases} Y_j, \qquad i = 1, \ldots, m; j = 1, \ldots, n.$$

The Mann-Whitney statistic U counts the number of times that Y-values are greater than X-values in the combined set of $N = m + n$ values. Since U is a linear function of W, the Wilcoxon test can be based on either the rank sum W or on the statistic U.

4. DISTRIBUTION-FREE TESTS

When X- and Y-observations come from the same population as specified by the two-sample hypothesis (1), the Y-ranks $R_1, \ldots, R_n$ constitute a random selection of n numbers from among the integers $1, \ldots, N$. As a result the exact sampling distribution of a test statistic like the Wilcoxon statistic that depends on the Y-ranks can be obtained by evaluating and tabulating its value for each of the $\binom{N}{n}$ equally likely selections of n ranks $R_1, \ldots, R_n$. We say that the test based on such a statistic is *distribution-free*, since an exact test can be carried out whatever the distribution $F(z)$ underlying the two-sample hypothesis (1), as long as $F(z)$ is continuous. In Section 8 we discuss the test modifications required by discontinuous data. As we shall see, the modified procedure is no longer distribution-free.

Many textbook authors use the terms "nonparametric" and "distribution-free" interchangeably. Such identification still further confuses the issue surrounding the term nonparametric. Nonparametric tests as they are customarily understood are distribution-free only when sampling from continuous populations and then only when the hypothesis being tested is true.

Tables for carrying out the Wilcoxon test are widely available (for example, [9]). Unless m and n are quite small, W is approximately normally distributed with mean $n(m + n + 1)/2$ and variance $mn(m + n + 1)/12$. The asymptotic normality follows from general theorems developed primarily for the study of nonparametric statistics. The reader is referred to Hájek [4] and Noether [14].

5. CONSISTENCY

We have introduced the Wilcoxon test as the rank analogue of the customary normal theory test involving a shift in location. We may ask whether the Wilcoxon test is appropriate for a class of alternatives that is wider than (4). This question is answered by establishing consistency conditions for the Wilcoxon test. A test is said to be *consistent* against a class of alternatives if, under the alternative, the probability of rejection converges to 1 as the sample size increases.

As before let X be a random variable with c.d.f. $F(z)$ and Y, a random variable with c.d.f. $G(z)$. We set

$$\pi = P(X < Y) = \int_{-\infty}^{\infty} F(z)\, dG(z). \tag{8}$$

Then from (7),

$$EU = \sum_i \sum_j EV_{ij} = \sum_i \sum_j P(X_i < Y_j) = mn\pi$$

and $p = U/mn$ is an unbiased estimate of π. It is not difficult to show that the variance of p converges to zero as both m and $n \to \infty$, so that p converges in probability to π. When the hypothesis (1) is correct, we have $\pi = 1/2$, and the Mann-Whitney test, and therefore the Wilcoxon test, is consistent against alternatives for which (8) differs from 1/2. This condition is satisfied for alternatives (4) with Δ different from 0. But it is also satisfied for the much larger class of alternatives (3). The Wilcoxon test is appropriate whenever we suspect that X-observations tend to be smaller than Y-observations, or vice versa.

6. POWER OF TEST

The probability $\pi = P(X < Y)$ plays an important role not only for questions of consistency, but also when investigating the power of the Wilcoxon test. The power of a test is the probability of rejecting the hypothesis being tested when it is false. Since the hypothesis is rejected whenever the test statistic takes a value in the critical region, exact power computations require the computation of probabilities associated with sample arrangements that result in sufficiently extreme values of the rank sum W. Such computations are complex and time consuming. For the case of normally distributed populations with identical variance σ^2 they were performed by Milton [12] for samples of size $\leqslant 7$ and selected values of the parameter $d = (\mu_y - \mu_x)/\sigma$.

Unless sample sizes m and n are quite small, approximate power computations are possible using the normal approximation. For the Mann-Whitney form of the test the appropriate mean is $\mu = mn\pi$. Corresponding computations (see, for example, Lehmann [11, p. 335]) give

$$\begin{aligned}\sigma^2 = mn\pi(1 - \pi) + mn(n - 1)(\pi' - \pi^2)\\ + nm(m - 1)(\pi'' - \pi^2)\end{aligned} \tag{9}$$

where $\pi' = P(X < Y_1 \text{ and } X < Y_2)$, $\pi'' = P(X_1 < Y \text{ and } X_2 < Y)$, and where X, X_1, X_2, Y, Y_1, Y_2 are mutually independent random variables having distributions $F(z)$ and $G(z)$, respectively. Under the null hypothesis $F(z) = G(z)$, $\pi' = \pi'' = 1/3$ and (9) reduces to $\sigma_0^2 = mn(m + n + 1)/12$, as stated earlier.

Against the alternative that Y-observations tend to be larger than X-observations, the hypothesis (1) is rejected at significance level α provided $U > \frac{1}{2}mn + z_\alpha\sigma_0$, where z_α is the upper α-percentile of the standard normal distribution. When $P(Y > X) = \pi$, we have approximately,

$$\text{power of test} = P(U > \tfrac{1}{2}mn + z_\alpha\sigma_0)$$
$$= P\left(\frac{U - mn\pi}{\sigma} > \frac{mn(\frac{1}{2} - \pi) + z_\alpha\sigma_0}{\sigma}\right).$$

Thus the power is given by the area under the standard normal curve to the right of

$$\frac{z_\alpha\sigma_0 - mn(\pi - \frac{1}{2})}{\sigma} \doteq z_\alpha - \frac{mn(\pi - \frac{1}{2})}{\sqrt{mn(m + n + 1)/12}},$$

where we have approximated σ by σ_0. To the extent that this approximation is valid, the power of the Wilcoxon test depends on

$$\pi - \frac{1}{2} = \int [F(z) - G(z)]\, dG(z).$$

The nonparametric parameter $\pi - 1/2$ seems a much more natural way to characterize alternatives than the noncentrality parameter $(\mu_y - \mu_x)^2/\sigma^2$ of normal theory.

7. A CONFIDENCE INTERVAL FOR Δ

We have seen that the Wilcoxon test based on the statistic W or equivalently, the Mann-Whitney test based on the statistic U can be used to test the hypothesis $\Delta = 0$, if the shift model (4) is appropriate. A confidence interval for the shift parameter Δ includes all values Δ_0 such that a test of the hypothesis $\Delta = \Delta_0$ against the two-sided alternative $\Delta \neq \Delta_0$ does not lead to rejection of the hypothesis. The hypothesis $\Delta = \Delta_0$ is tested by applying the Wilcoxon or Mann-Whitney test to observations

$$X'_i = X_i + \Delta_0 \quad \text{and} \quad Y'_j = Y_j.$$

The following general method then provides upper and lower confidence bounds for Δ.

Let θ be a parameter for which we want to find a confidence interval. We assume that the hypothesis $\theta = \theta_0$ against the two-sided alternative $\theta \neq \theta_0$ can be tested as follows: We compute two test statistics $T' = \#(u\text{'s} < \theta_0)$ and $T = \#(u\text{'s} > \theta_0)$, where $u_1, \ldots, u_M$ are appropriate sample quantities, and reject the hypothesis $\theta = \theta_0$ at significance level α provided the smaller of T and T' is $\leqslant c$. The confidence interval with confidence coefficient $\gamma = 1 - \alpha$ is then bounded by the $(c + 1)$st smallest and largest of the u's,

$$u_{(c+1)} < \theta < u_{(M-c)}. \tag{10}$$

The inequality (10) follows, since the hypothesis $\theta = \theta_0$ is rejected whenever $u_{(c+1)} \geqslant \theta_0$ or $u_{(M-c)} \leqslant \theta_0$.

In the case of the shift parameter Δ, we use as the statistic T, the Mann-Whitney statistic

$$U = \#(X_i' < Y_j') = \#(X_i + \Delta_0 < Y_j) = \#(Y_j - X_i > \Delta_0);$$

as the statistic T', we use

$$U' = \#(X_i' > Y_j') = \#(X_i + \Delta > Y_j) = \#(Y_j - X_i < \Delta_0).$$

Thus, for the shift parameter Δ the u's are the mn differences $Y_j - X_i$ of all possible pairs of X- and Y-observations. The value c is the lower critical value of the two-sided Mann-Whitney test with significance level $\alpha = 1 - \gamma$. Since the Mann-Whitney test is distribution-free, so is the resulting confidence interval. The confidence statement (10) holds with probability $\gamma = 1 - \alpha$ irrespective of the distribution $F(z)$ in model (4), as long as $F(z)$ is continuous.

We shall call the set of differences $u_{ij} = Y_j - X_i$ the set of "elementary estimates" for the shift parameter Δ. The class of confidence intervals resulting from the Wilcoxon test consists of intervals that are bounded by symmetrically located elements in the ordered array of all mn elementary estimates $Y_j - X_i$. Hodges and Lehmann [7] have shown that the median of all mn elementary estimates furnishes a point estimate of Δ which has desirable properties.

8. TIED OBSERVATIONS

So far we have assumed that distributions $F(z)$ and $G(z)$ are continuous. As a result we were able to ignore difficulties arising from possible ties among the observations. In actual applications ties among the observations do occur either as the result of discontinuities in the population distribution or insufficient precision of the measurements. We now investigate how to deal with ties.

We first consider confidence intervals. Population discontinuities will in general result in ties among the elementary estimates $u_{ij} = Y_j - X_i$. This does not affect our choice of confidence bounds. We can still consider confidence intervals bounded by the $(c + 1)$st smallest and largest of the u_{ij}'s. What is affected by population discontinuities is the confidence coefficient associated with a given value c. The derivation of the interval (10) leads to an open interval. For continuous populations the probability is zero that a confidence bound $Y_j - X_i$ coincides with the true parameter value Δ. Thus, in the continuous case, instead of the open interval (10), we can use the closed interval

$$u_{(c+1)} \leqslant \Delta \leqslant u_{(M-c)} \tag{11}$$

without changing the confidence coefficient γ. The discrete case is different. Now a confidence bound may coincide with the true value Δ with positive probability. In that case the open interval (10) provides incorrect information while the closed interval (11) provides correct information. The confidence coefficient γ_0 associated with the open interval may be smaller than the confidence coefficient γ_c associated with the closed interval. It is shown in [13] that $\gamma_0 \leqslant \gamma \leqslant \gamma_c$, where γ is the distribution-free confidence coefficient of the continuous case associated with the value c in (10) and (11). Thus, it is advisable to use the closed interval in practice. In case of continuous populations, it provides an exact interval. In case of discrete populations, it provides a conservative interval (it covers the true value Δ with probability at least as great as claimed by the nominal confidence coefficient γ).

The case of hypothesis testing is more complex. In the presence of ties, the test statistic requires redefinition. Several different approaches are possible. In this brief account we consider only the one based on midranks, which assigns to tied observations the average of the ranks for which the observations are tied. The redefined test statistic W is the sum of the midranks associated with Y-observations.

While we have solved the problem of how to compute the test statistic, we have raised a ew and more serious problem. What is the distribution of edefined test statistic under the null hypoth- sis? We can use the ame approach as in the untied case. When the null hypothesis is true, the $\binom{m+n}{n}$ possible assignments of the *observed* midranks to X- and Y-observations are equally likely, and at least theoretically we can obtain the exact distribution of W by enumeration. This result reveals two aspects of the Wilcoxon test—and of nonparametric tests in general. For discrete populations the tests are *conditional* tests depending on the ties in the observed samples. Further, the tests are no longer distribution-free, but depend on the usually unknown population discontinuities that produce the observed ties among the observations. For practical purposes, these complications are rarely important. In the case of the Wilcoxon test, unless m and n are quite small, we can still use the normal approximation to the exact distribution of W. As in the distribution-free case, $EW = n(N + 1)/2$, but

$$\text{Var } W = \frac{1}{12} mn(m + n + 1)\left[1 - \frac{\sum (t^3 - t)}{N^3 - N}\right],$$

where summation extends over all sample ties and t denotes the extent of any one tie. (For untied observations we set $t = 1$.) The expression in square brackets—the correction for ties—equals 1 in the untied case; it is less than 1 if ties are present. Computations show that the correction factor can for most practical purposes be ignored, unless a small number of ties involve a large proportion of the observations. For example, if we have 20 observations consisting of 10 ties of size 2 each, the correction factor equals $1 - 10(8 - 2)/(8000 - 20) = .992$ while for one tie of size 10 and 10 untied observations we have $1 - (1000 - 10)/(8000 - 20) = .876$. Referring the test statistic to standard tables results in a conservative test, that is, a test whose true significance level is at most as large as the nominal level indicated by the table.

9. EFFICIENCY

Nonparametric tests like the Wilcoxon test use only information about the relative position of X- and Y-observations, while normal-theory procedures depend on the actual sample observations. As a consequence, it is often claimed that nonparametric procedures are "wasteful" of information. To judge the relevance of such claims we shall now compare the Wilcoxon test and the confidence interval and point estimate derived from the Wilcoxon test with the corresponding procedures suggested by normal theory. The normal theory point estimate of the shift parameter Δ is the difference of sample means $\bar{Y} - \bar{X}$, while tests of hypotheses and confidence intervals use the well-known two-sample t-statistic based on $\bar{Y} - \bar{X}$. Two statistical procedures for solving the same inference problem are often compared in terms of *relative efficiency*. We say that procedure 1 has efficiency e relative to procedure 2 provided procedure 1 based on N_1 observations is equivalent to procedure 2 based on $N_2 = eN_1$ observations. Thus if $e = .80$, method 2 will do as well as method 1 with 20 percent fewer observations. Equivalence means different things depending on whether we want to test a hypothesis, find a confidence interval or find a point estimate. We shall have the following interpretations in mind. Two test procedures are equivalent if for the same significance level both have the same power with respect to a given alternative. Two confidence interval procedures are equivalent if they produce intervals of equal length. Two procedures for finding point estimates are equivalent if they produce estimates of equal accuracy. There are many theoretical and practical complications associated with efficiency. But these need not concern us in this paper. For our discussion of the efficiency of the Wilcoxon test and the related estimation procedures, we consider the model $G(z) = F(z - \Delta)$ according to which Y-observations have been shifted relative to X-observations by an amount Δ. We compare the Wilcoxon test for testing the hypothesis $\Delta = 0$ with the corresponding two-sample t-test based on the statistic $t = (\bar{y} - \bar{x})/s\sqrt{1/m + 1/n}$, where s is the pooled estimate of the population standard deviation. Most discussions of the efficiency of nonparametric methods restrict comparisons to the case of normally distributed populations. However, in practice, normal theory methods are often used when there is no assurance that the assumption of normally distributed observations is even approximately satisfied. It is then of considerable practical and theoretical interest to study efficiency properties of nonparametric procedures for models involving other than normal distributions. For the two-sample shift model the efficiency of the Wilcoxon test relative to the t-test is found to be

$$e = 12\left[\sigma \int f^2(z)\, dz\right]^2, \tag{12}$$

where $f(z) = F'(z)$ is the density function and σ, the population standard deviation (see, for example, Noether [13]). It has been shown by Hodges and Lehmann [6] that $e \geqslant .864$, whatever the distribution $F(z)$. For normal distributions, $e = 3/\pi = .955$. For the logistic distribution with density

$$f(z) = \frac{\exp\{-(z - \eta)/\theta\}}{\theta[1 + \exp\{-(z - \eta)/\theta\}]^2},$$

which is often compared to the normal distribution, we find $e = (\pi/3)^2 = 1.10$ and for the double exponential distribution with density

$$f(z) = \tfrac{1}{2}\theta e^{-\theta|z-\eta|}$$

we find $e = 3/2$. For distributions without finite standard deviation like the Cauchy distribution, we have $e = \infty$, so that in this case the Wilcoxon test is infinitely better than the t-test. Using the Wilcoxon test in place of the t-test rarely leads to a substantial "waste" of observations, but may result in a substantial gain, as in the case of distributions with "heavy" tails such as the exponential and Cauchy distributions.

We have stated efficiency results using the terminology of hypothesis testing. Hodges and Lehmann [7] and Lehmann [10] have shown that the same efficiency results hold when we estimate the shift parameter Δ by means of the confidence interval or point estimate discussed in Section 7.

10. LINEAR RANK TESTS FOR THE TWO-SAMPLE PROBLEM

In Section 3 we noted that the Wilcoxon test belonged to the class of linear rank tests. We now define linear rank tests in general.

Let $a = \{a_1, \ldots, a_N\}$ be a set of scores with $a_1 \leqslant \cdots \leqslant a_N$. In generalization of the Wilcoxon statistic W we consider tests based on statistics

$$W_{[a]} = \sum_{j=1}^{n} a(R_j),$$

where $a(k) = a_k$ and $R_j = r[Y_j]$. For scores $a_k = k$, the $W_{[a]}$-test becomes the Wilcoxon test.

Another well-known test, the median test, has scores

$$a_1 = \cdots = a_r = 0, \qquad a_{r+1} = \cdots = a_N = 1, \tag{13}$$

where $r = N/2$, if N is even, and $r = (N+1)/2$, if N is odd. The name, median test, refers to the fact that for the scores (13), $W_{[a]}$ is equal to the number of Y-observations that surpass the median for the combined X- and Y-samples.

The exact distribution of $W_{[a]}$ when the two-sample hypothesis (1) is true is obtained by enumeration as in the case of the Wilcoxon statistic. Actually, unless m and n are quite small, under very general conditions on the scores a_k, $W_{[a]}$ is approximately normally distributed with $EW_{[a]} = n\bar{a}$ and

$$\operatorname{Var} W_{[a]} = \frac{mn}{N(N-1)} \sum (a - \bar{a})^2,$$

where $\bar{a} = \sum a/N$.

We consider again the shift model $G(z) = F(z - \Delta)$. If information about the distribution type $F(z)$ is available, it is possible to select scores a_k which maximize the power of the corresponding linear rank test for the particular type of distribution. We define the score function

$$\phi(u, f) = -f'(F^{-1}(u))/f(F^{-1}(u)), \qquad 0 < u < 1,$$

where $f(z) = F'(z)$. One way to obtain scores that are (asymptotically) optimal is to use

$$a_k = \phi\left(\frac{k}{N+1}, f\right), \qquad k = 1, \ldots, N \tag{14}$$

or linear functions thereof (see Hájek [5], Chapter 8).

For the Wilcoxon scores $a_k = k$, $\phi(u,f)$ must be a linear function of u. This occurs for the logistic distribution with density $f(z) = e^{-z}/(1 + e^{-z})^2$, for which $\phi(u,f) = 2u - 1$. The scores for the median test correspond to the double exponential distribution with density $f(z) = e^{-|z|}/2$ and score function $\phi(u,f) = \text{sign}(2u - 1)$.

For the normal density $f(z) = e^{-z^2/2}/\sqrt{2\pi}$, we find $-f'(z)/f(z) = z$, so that $\phi(u,f) = \Phi^{-1}(u)$, where $\Phi(z)$ is the standard normal cumulative distribution function. Thus the a_k given by (14) divide the area under the normal density function into $N + 1$ equal parts. These scores were first suggested by van der Waerden. Chernoff and Savage [2] have shown that the corresponding linear rank test has asymptotic efficiency 1 relative to the t-test, if X- and Y-observations are indeed normally distributed and greater than 1, for any other distribution.

In practical applications, the appropriate distribution type $F(z)$ that determines the optimal scores (14) is rarely known. To get around this difficulty statisticians have begun to develop so-called *adaptive procedures* (Hogg [8]).

The basic rationale behind the adaptive approach emerges if we compare the three linear rank tests that we have just discussed. Linear rank tests differ with respect to the amount of weight that they assign to extreme observations. (The comparison becomes clearer, if scores a_k are replaced by $a_k' = a_k - \bar{a}$, so that all tests have expected value 0.) The median test assigns the same weight to extreme observations as to central observations. Thus for populations with "heavy" tails extreme observations are no more indicative of a shift in location than are central observations. The Wilcoxon test, which is most appropriate for populations with tails that are somewhat heavier than the tails of a normal distribution, puts some emphasis on extreme observations by using weights that are proportional to the rank number of an observation. Finally, the van der Waerden test, which is appropriate for normally distributed observations, uses weights that are essentially proportional to the observations (measured in standard units).

Our examples show that the appropriate choice of test statistic $W_{[a]}$ depends largely on the tail behavior of the distribution type $F(z)$. Adaptive procedures use the X- and Y-observations to estimate tail weight and then select a test statistic that is appropriate for the indicated tail weight.

All $W_{[a]}$-tests can be used to find confidence intervals for the shift parameter Δ. These intervals are again bounded by elementary estimates $u_{ij} = Y_j - X_i$, the choice depending on weights associated with each u_{ij} as described in Bauer [1]. By choosing an appropriate $W_{[a]}$-test, we can affect the length of the resulting confidence interval.

11. THE KOLMOGOROV-SMIRNOV TEST

One of the oldest and also best-known two-sample tests is the Kolmogorov-Smirnov test based on the statistic

$$D = \max_z |F_m(z) - G_n(z)|,$$

where $F_m(z) = \#(X_i \leqslant z)/m$ and $G_n(z) = \#(Y_j \leqslant z)/n$ are the sample (or empirical) distribution functions for the X- and Y-samples, respectively. The value of the statistic D depends only on the relative position of X- and Y-observations, so that theoretically the exact distribution of D under the two-sample hypothesis can be obtained by evaluating and tabulating D for each of the $\binom{m+n}{n}$ equally likely sample arrangements. Mathematically more attractive methods for finding the distribution of D make use of random walk models.

The Kolmogorov-Smirnov test rejects the two-sample hypothesis, when D is sufficiently large. Now assume that for $z = z_0$, $G(z_0) \neq F(z_0)$. According to the law of large numbers, $G_n(z_0)$ converges in probability to $G(z_0)$ and $F_m(z_0)$ converges to $F(z_0)$. With increasing sample sizes we are sure to reject the hypothesis $G(z) = F(z)$, when it is false. The Kolmogorov-Smirnov test has the interesting property of being consistent against the completely general alternative (2), $G(z) \neq F(z)$. While this may look like a very attractive and desirable property, it has its disadvantages. If the experimenter has some limited alternative like stochastic ordering in mind, he will usually be able to achieve greater power with an appropriately chosen linear rank test.

12. THE ONE-SAMPLE LOCATION PROBLEM

Consider observations

$$Z_k = \eta + e_k, \qquad k = 1, \ldots, N, \tag{15}$$

where η is an unknown location parameter and the e_k are independently and identically distributed random errors. We assume that the error distribution is symmetric about zero. The one-sample location problem is concerned with the estimation of η and tests of hypotheses involving η. We start our discussion with tests of hypotheses.

For model (15), the following device allows us to convert a two-sample test into a test of the hypothesis $\eta = \eta_0$. For simplicity of notation we assume for the moment that the Z_k have been reordered according to size and that

$$Z_1 < \cdots < Z_m < \eta_0 < Z_{m+1} < \cdots < Z_N.$$

Let

$$\begin{aligned} X_i &= |Z_i - \eta_0|, & i &= 1, \ldots, m \\ Y_j &= Z_{m+j} - \eta_0, & j &= 1, \ldots, n = N - m. \end{aligned}$$

If the hypothesis $\eta = \eta_0$ is correct, X- and Y-observations are identically distributed in view of the assumed symmetry about zero of the errors e_k. Any two-sample test applied to the X's and Y's then becomes a test of the hypothesis of symmetry of the Z's about η_0. In particular, let $a = \{a_1, \ldots, a_N\}$ be a set of scores satisfying $0 \leqslant a_1 \leqslant \cdots \leqslant a_N$. The corresponding linear rank test of the hypothesis $\eta = \eta_0$ is based on the test statistic

$$T_{[a]} = \sum_{Z_k > \eta_0} a(R_k)$$

where $a(k) = a_k$ and $R_k = r[|Z_k - \eta_0|]$.

While the device just described furnishes test statistics for testing the one-sample hypothesis, it is important to realize that the sampling distribution of the test statistic in the one-sample case differs from the sampling distribution of the corresponding test statistic in the two-sample case. In the two-sample case, the sample sizes m and n

are constants. In the one-sample case, they are variables, since $n = \#(Z_k > \eta_0)$ depends on the sample observations.

The sampling distribution of $T_{[a]}$ is found as follows. Let $r[|Z_k - \eta_0|] = h$. Then we can write

$$T_{[a]} = \sum_{h=1}^{N} a_h t_h, \qquad t_h = \begin{cases} 1 \\ 0 \end{cases} \text{if } Z_k \begin{cases} > \\ < \end{cases} \eta_0. \tag{16}$$

If the Z_k are symmetrically distributed about η_0, $P(t_h = 0) = P(t_h = 1) = 1/2$ and t_g and t_h are independent for $g \neq h$. The exact distribution of $T_{[a]}$ is obtained by finding the value of $T_{[a]}$ for each of the 2^N equally likely assignments of zero's and one's in (16). General central limit theorems for sums of independent (though not identically distributed) components show that under weak restrictions on the scores a_k, $T_{[a]}$ has an asymptotic normal distribution as $N \to \infty$. The appropriate mean and variance are

$$ET_{[a]} = \frac{1}{2}\sum_{h=1}^{N} a_h \qquad \text{and} \qquad \operatorname{var} T_{[a]} = \frac{1}{4}\sum_{h=1}^{N} a_k^2.$$

13. THE WILCOXON ONE-SAMPLE TEST

For scores $a_k = k$, the two-sample Wilcoxon test becomes the one-sample Wilcoxon or signed rank test based on the statistic

$$T = \sum_{Z_k > \eta_0} r[|Z_k - \eta_0|].$$

Exact tables for carrying out the Wilcoxon test are widely available (for example, [9]). When using the normal approximation, we have $\mu = N(N + 1)/4$ and $\sigma^2 = N(N + 1)(2N + 1)/24$.

It is interesting to examine consistency conditions for the Wilcoxon one-sample test. It can be shown (see, for example, Noether [13], p. 47) that the test is consistent against alternatives for which $|\epsilon pq + p - \frac{1}{2}| \neq 0$. Here the quantities ϵ, p and q have the following meaning. For simplicity we assume that $\eta_0 = 0$. We define random variables X and Y which are distributed like absolute values of negative Z-observations and positive Z-observations respectively. Then $\epsilon = P(X < Y) - P(X > Y)$, $p = P(Z > 0)$, $q = 1 - p$. It can be shown that if the distribution of Z is symmetric about $\eta \neq 0$, then $(\epsilon pq + p - \frac{1}{2})\eta > 0$, so that the test is consistent against alternatives according to which the distribution is symmetric about $\eta \neq \eta_0$. Additionally the test is consistent against alternatives for which $p = 1/2$ and $\epsilon \neq 0$. This condition implies that Z has median 0, but that the distribution of Z is not symmetric. The Wilcoxon test is a test for location only if the symmetry assumption is appropriate. Extreme values of the test statistic may also be an indication of asymmetry about the hypothesized central value.

If the error distribution in (15) has density $f(z)$, the efficiency of the Wilcoxon one-sample test relative to the one-sample t-test is given by (12).

The problem of finding confidence intervals and point estimates for the center η of a symmetric distribution can be solved analogously to the problem of estimating the shift parameter Δ in the two-sample problem. The elementary estimates of η are

$$u_{ij} = \tfrac{1}{2}(Z_i + Z_j), \qquad 1 \leqslant i \leqslant j \leqslant N.$$

The median of all u_{ij}'s furnishes a point estimate of η, while confidence intervals corresponding to the various linear rank tests are bounded by elements of the ordered array of u_{ij}'s. In particular, the confidence interval derived from the Wilcoxon one-sample test is bounded by the $(c + 1)$st smallest and largest u_{ij}, where c is the lower critical value of the test.

14. THE SIGN TEST

Another linear rank test has considerable practical importance. For scores $a_1 = \cdots = a_N = 1$, $T_{[a]}$ equals the number of positive differences $Z_k - \eta_0$. This test is known as the sign test. Its sampling distribution is the binomial distribution for N trials and success probability $p = P(Z > \eta_0)$. In particular, if η_0 is the population median, $p = 1/2$. The sign test does not require that the population be symmetric, only that η be the population median. The test is consistent against the alternative that the true population median differs from the hypothetical median η_0. For estimation purposes, the relevant u_{ij}'s are the ones with $i = j$, that is, the observations themselves.

Compared to the one-sample t-test, the relative efficiency of the sign test is given by $4\sigma^2 f^2(0)$. For the family of normal distributions, e has the value $2/\pi = .64$, while for the family of double exponential distributions, $e = 2$. As a test of location the sign test is particularly useful for distributions with "heavy" tails.

We can also compare the sign test with the Wilcoxon one-sample test. The efficiency of the sign test relative to the Wilcoxon test is $f^2(0)/3[\int f^2(z)\,dz]^2$ which equals 2/3 for normal distributions and 4/3 for double exponential distributions.

15. THE KOLMOGOROV STATISTIC

The sample distribution function which we introduced in connection with the Kolmogorov-Smirnov test has important applications for the one-sample case. Let $Z_1, \ldots, Z_N$ constitute a random sample from a population with cdf $F(z)$. For $-\infty < z < \infty$, the sample c.d.f. is given by $F_N(z) = \#(Z_k \leqslant z)/N$, the proportion of observations that have values $\leqslant z$. As we have already observed, as $N \to \infty$, $F_N(z)$ converges in probability to $F(z)$ for every z. Actually, a considerably stronger statement is possible. According to the Glivenko-Cantelli Theorem (see, for example, Fisz [3]), we have

$$\sup_z |F_N(z) - F(z)| \to 0 \text{ with probability 1.}$$

The theorem has important statistical implications. $F_N(z)$, which is a step function that increases by $1/N$ at each observation Z_k, provides an estimate of $F(z)$. The distance

$$K = \sup_z |F_n(z) - F(z)|$$

between $F(z)$ and its estimate is known as the Kolmogorov statistic. Its usefulness for statistical purposes derives from the fact that if $F(z)$ is continuous the distribution of K depends only on N but not on the distribution $F(z)$. The distribution-free nature of K follows by observing that the transformation $U = F(Z)$ transforms the

Z-sample into a sample from the uniform distribution on the interval (0, 1) without changing the value of K whatever the (continuous) distribution $F(z)$.

Let K_α satisfy $P(K \leqslant K_\alpha) = 1 - \alpha$. A distribution-free test of the hypothesis $F(z) = F_0(z)$, where $F_0(z)$ is a given cdf, can then be based on the Kolmogorov test which rejects the hypothesis at significance level α provided

$$K = \sup_z |F_N(z) - F_0(z)| > K_\alpha.$$

This test is consistent against alternatives $F(z) \neq F_0(z)$. Alternatively, we can rewrite $K \leqslant K_\alpha$ as

$$L(z) \leqslant F(z) \leqslant U(z), \qquad -\infty < z < \infty$$

where $L(z) = \max(0, F_N(z) - K_\alpha)$ and $U(z) = \min(1, F_N(z) + K_\alpha)$, to form a confidence band for the unknown cdf $F(z)$ which has confidence coefficient $1 - \gamma$.

It can be shown that for discontinuous distributions $F(z)$, $P(K \leqslant K_\alpha) \geqslant 1 - \alpha$, so that for discontinuous distributions the Kolmogorov test and the related confidence band are conservative.

16. SOME c-SAMPLE PROBLEMS

We have seen that two-sample tests suggest test procedures for the one-sample symmetry problem. Two-sample tests also suggest possible approaches to the c-sample problem.

In its simplest form the c-sample problem is formulated as follows. For $g = 1, \ldots, c$, we have independent random samples $X_{1g}, \ldots, X_{n_g g}$ and want to test the hypothesis that all $N = n_1 + \cdots + n_c$ observations have come from one and the same population. Often, the c samples correspond to c "treatments." The hypothesis then implies that treatments do not differ in their effectiveness.

As in the two-sample case we assign ranks $R_{ig} = r[X_{ig}]$ to the N observations and, for scores $a_1 \leqslant \cdots \leqslant a_N$, consider sums of scores

$$W_g = \sum_{i=1}^{n_g} a(R_{ig}).$$

The decision of whether or not to accept the hypothesis under consideration is then based on some suitable function of the W_g such as $\sum_{g=1}^{c} d_g(W_g - n_g\bar{a})^2$, where the d_g are weights and large values of the statistic indicate rejection of the hypothesis.

General limit theorems show that with suitable restrictions on the scores a_k,

$$Q = (N-1)\left[\sum_{k=1}^{N}(a_k - \bar{a})^2\right]^{-1} \sum_{g=1}^{c} \frac{1}{n_g}(W_g - n_g\bar{a})^2$$

has approximately a chi-square distribution with $c - 1$ degrees of freedom. For scores $a_k = k$, we obtain the Kruskal-Wallis test based on the statistic

$$H = \frac{12}{N(N+1)} \sum_{g=1}^{c} \frac{1}{n_g}\left(R_{+g} - \frac{n_g(N+1)}{2}\right)^2,$$

where

$$R_{+g} = \sum_{i=1}^{n_g} R_{ig}$$

is the sum of the ranks for the gth sample.

Experimental conditions often dictate that observations be collected from different blocks. Observations are then ranked separately within each block and the W_g are computed by summing over blocks.

The simplest case arises when each of c treatments is used exactly once in each block and we use scores $a_k = k$. The test statistic then is the Friedman statistic

$$\chi_r^2 = \frac{12}{mc(c+1)} \sum_{g=1}^{c} [R_{+g} - \tfrac{1}{2}m(c+1)]^2$$

where R_{+g} is the sum of the ranks assigned to treatment g in each of m blocks.

The problem underlying the Friedman test is also known as the problem of m rankings. In this problem, we assume that m "judges" (the blocks) rank c "contestants" (the treatments) in order of preference. The purpose of such an experiment is to find out whether there is some agreement among the m judges with respect to their order of preference. Agreement is indicated by a large value of χ_r^2. This corresponds to the existence of an objective ordering of the treatments in the c-sample problem. A low value of χ_r^2 may be indicative of random arrangement of the objects by the judges due to a lack of preference. This corresponds to the absence of any objective ordering of the treatments in the c-sample problem. On the other hand, a low value of χ_r^2 may also arise from definite disagreements among the judges. The following example of the rankings that four judges assigned to five contestants reveals considerable agreement

Judges	Contestants A	B	C	D	E
1	1	2	3	4	5
2	2	1	4	3	5
3	4	5	3	1	2
4	5	3	4	2	1
R_{+g}	12	11	14	10	13

between judges 1 and 2 and between judges 3 and 4, but strong disagreement between the two groups of judges. The value of χ_r^2 is 1.

17. A TEST OF RANDOMNESS

The hypotheses underlying the one-, two-, and c-sample problems are special cases of the following hypothesis of randomness. We have N independent random variables $Z_1, \ldots, Z_N$ with cumulative distribution functions $F_1(z), \ldots, F_N(z)$. The hypothesis of randomness states that

$$F_1(z) = \cdots = F_N(z). \tag{17}$$

This hypothesis reduces to the c-sample hypothesis if we set $N = n_1 + \cdots + n_c$ and assume that

$$F_{n_1+\cdots+n_{g-1}+1}(z) = \cdots = F_{n_1+\cdots+n_g}(z), \qquad g = 1, \ldots, c.$$

For $c = 2$, this is the two-sample setup. In the one-sample case we usually specify certain definite characteristics of the common distribution (17).

Possible alternatives to the hypothesis of randomness are manifold. We consider the stochastic ordering alternative

$$F_1(z) \geqslant F_2(z) \geqslant \cdots \geqslant F_N(z),$$

which represents a monotone (upward) trend among the Z_k. A test of randomness against the alternative of a monotone trend can be based on Kendall S,

$$S = \#(Z_i < Z_j) - \#(Z_i > Z_j), \qquad 1 \leqslant i < j \leqslant N.$$

Values of S near zero are indicative of randomness; sufficiently large positive values suggest an upward trend, while sufficiently large negative values suggest a downward trend. As in previous cases, the exact distribution of S (assuming randomness) is obtained by tabulating S for the $N!$ equally likely rank arrangements of the Z_k. Unless N is quite small, a normal approximation with $\mu = 0$ and $\sigma^2 = N(N-1)(2N+5)/18$ can be used.

18. REGRESSION

We assume that the Z_k satisfy the following regression model,

$$Z_k = \eta + \Delta(t_k) + e_k, \tag{18}$$

where η is an unknown constant, $\Delta(t)$ is some unknown function, the regression constants t_k are known numbers, and the e_k are random errors assumed to be independently and identically distributed with median 0. If $\Delta(t) = 0$ for all t, the variables Z_k satisfy the hypothesis of randomness, so that the hypothesis that the Z_k do not depend on the regression variable t can be tested by means of a test of randomness. In particular, a test against the alternative that $\Delta(t)$ is a monotone function can be based on a modified version of Kendall S,

$$S = \#(Z_i < Z_j) - \#(Z_i > Z_j), \qquad t_1 \leqslant t_i < t_j \leqslant t_N. \tag{19}$$

If in (18), $\Delta(t) = \Delta \cdot t$, where Δ is a constant, we have the case of linear regression,

$$Z_k = \eta + \Delta \cdot t_k + e_k. \tag{20}$$

The regression parameter Δ in (20) is estimated as follows. We find the elementary estimates

$$u_{ij} = \frac{Z_j - Z_i}{t_j - t_i}, \qquad t_1 \leqslant t_i < t_j \leqslant t_N. \tag{21}$$

The median of the u_{ij}'s provides a point estimate of Δ, while a confidence interval is bounded by two elements among the ordered u_{ij}'s determined by the test based on (19).

For the choice $t_i = 0$, $i = 1, \ldots, m$; $t_{m+j} = 1$, $j = 1, \ldots, N - m = n$, the linear regression model (20) reduces to the two-sample shift model. If we write $Z_i = X_i$, $i = 1, \ldots, m$; $Z_{m+j} = Y_j$, $j = 1, \ldots, n$, the statistic S in (19) becomes

$$S = \#(X_i < Y_j) - \#(X_i > Y_j) = U - U',$$

where U is the Mann-Whitney statistic (7) and $U' = \#(X_i > Y_j)$ is an alternative form. The test based on S when applied to two-sample data is equivalent to the Wilcoxon-Mann-Whitney test. Correspondingly the elementary estimates (21) for the regression problem reduce to the elementary estimates $u_{ij} = Y_j - X_i$ for the two-sample shift parameter.

19. RANK CORRELATION

The regression model (18) assumes that an investigator has selected regression constants $t_1, \ldots, t_N$ at which he wants to observe a variable Y in order to study how Y changes with t. If the investigator wants to study the mutual relationship of two random variables T and Z and has N pairs of observations $(T_1, Z_1), \ldots, (T_N, Z_N)$, Kendall S as given by (19) provides a test of the hypothesis that the random variables T and Z are independent against the alternative that, say, Z tends to increase (decrease) as T increases. Under the hypothesis of independence each of the possible rank arrangements of $Z_1, \ldots, Z_N$ is equally likely, whatever the rank arrangement of $T_1, \ldots, T_N$ that has been observed, and S has the same distribution as before.

Let (T_1, Z_1) and (T_2, Z_2) be two independent sets of random variables, each set distributed like the pair (T, Z). To study the strength of relationship between T and Z, we define two probabilities π_c and π_d, called the probabilities of concordance and discordance,

$$\pi_c = P\{(T_1 - T_2)(Z_1 - Z_2) > 0\}$$

and

$$\pi_d = P\{(T_1 - T_2)(Z_1 - Z_2) < 0\}.$$

The quantity

$$\tau = \pi_c - \pi_d$$

is known as the Kendall rank correlation coefficient. Clearly

$-1 \leqslant \tau \leqslant 1$ with the extremes being attained if either π_c or π_d equals 1. If T and Z are independent, $\tau = 0$. But the reverse is not true. Since S can be computed as

$$S = \sum_{i<j} \operatorname{sign}[(T_j - T_i)(Z_j - Z_i)],$$

$t = 2S/N(N-1)$ is an estimate of τ.

20. LINEAR RANK STATISTICS

Given N pairs of observations (T_k, Z_k), we set $R_k = r[Z_k]$, $Q_k = r[T_k]$ and consider tests of independence based on linear rank statistics

$$S_{[a,b]} = \sum_{k=1}^{N} a(R_k)b(Q_k), \tag{22}$$

where $a_1 \leqslant \cdots \leqslant a_N$ and $b_1 \leqslant \cdots \leqslant b_N$ are sets of scores. We note that Kendall S is not a linear rank statistic. The exact distribution of $S_{[a,b]}$ is obtained by enumeration over equally likely rank arrangements. The mean and variance for the normal approximation are

$$ES_{[a,b]} = N\bar{a}\bar{b}$$

and

$$\operatorname{var} S_{[a,b]} = \sum (a - \bar{a})^2 \sum (b - \bar{b})^2/(N-1).$$

If in (22) we choose $a_k = b_k = k$, we obtain the Spearman rank test based on the statistic $\sum R_k Q_k$. Standardization gives the Spearman rank correlation coefficient

$$r_S = \frac{12 \sum R_k Q_k}{N^3 - N} - \frac{3(N+1)}{N-1},$$

which can also be computed as the regular correlation coefficient r applied to the rank pairs (R_k, Q_k). r_S has maximum value 1 when $R_k = Q_k$ and minimum value -1 when $R_k + Q_k = N + 1$, the same as the estimate t of the Kendall rank correlation coefficient τ.

To ob ain linear rank tests for the hypothesis $\Delta = 0$ in (20), we set $b_k = t_k$, giving linear rank statistics

$$S_{[a]} = \sum_{k=1}^{N} t_k a(R_k). \tag{23}$$

To apply (23) to the two-sample problem, we choose again

$t_1 = \cdots = t_m = 0$, $t_{m+1} = \cdots = t_{m+n} = 1$ and write $Z_i = X_i$, $i = 1, \ldots, m$; $Z_{m+j} = Y_j$, $j = 1, \ldots, n$. Then $S_{[a]} = \sum_{j=1}^{n} a(R_j) = W_{[a]}$ in the notation of Section 10. In particular, for scores $a_k = k$, we have again the Wilcoxon statistic $W = \sum_{j=1}^{n} r[Y_j]$.

BIBLIOGRAPHY

1. D. F. Bauer, "Constructing confidence sets using rank statistics," *J. Amer. Statist. Assoc.*, **67** (1972), 687–690.
2. H. Chernoff and I. R. Savage, "Asymptotic normality and efficiency of certain nonparametric test statistics," *Ann. Math. Statist.*, **29** (1958), 972–994.
3. M. Fisz, *Theory of Probability and Mathematical Statistics*, Wiley, New York, 1963.
4. J. Hájek, "Some extensions of the Wald-Wolfowitz-Noether Theorem," *Ann. Math. Statist.*, **32** (1961), 506–523.
5. ———, *A Course in Nonparametric Statistics*, Holden-Day, San Francisco, 1969.
6. J. L. Hodges, Jr., and E. L. Lehmann, "The efficiency of some nonparametric competitors of the *t*-test," *Ann. Math. Statist.*, **27** (1956), 324–335.
7. ———, "Estimates of locations based on rank tests," *Ann. Math. Statist.*, **34** (1963), 598–611.
8. R. V. Hogg, "Adaptive robust procedures: a partial review and some suggestions for future applications and theory," *J. Amer. Statist. Assoc.*, **69** (1974), 909–927.
9. Institute of Mathematical Statistics, *Selected Tables in Mathematical Statistics*, vol. 1, Amer. Math. Soc., 1973.
10. E. L. Lehmann, "Nonparametric confidence intervals for a shift parameter," *Ann. Math. Statist.*, **34** (1963), 1507–1512.
11. ———, *Nonparametrics; Statistical Methods Based on Ranks*, Holden-Day, San Francisco, 1975.
12. R. C. Milton, *Rank Order Probabilities*, Wiley, New York, 1970.
13. G. E. Noether, *Elements of Nonparametric Statistics*, Wiley, New York, 1967.
14. ———, "A central limit theorem with nonparametric applications," *Ann. Math. Statist.*, **41** (1970), 1753–1755.
15. J. Wolfowitz, "Additive partition functions and a class of statistical hypotheses," *Ann. Math. Statist.*, **13** (1942), 247–279.

Readers may also want to consult the following general reference:

M. Hollander and D. A. Wolfe, *Nonparametric Statistical Methods*, Wiley, New York, 1973.

CHI-SQUARE TESTS*

David S. Moore

1. INTRODUCTION

Statistics is the science of collecting, describing and interpreting data. The most common mathematical model underlying statistical interpretation of data assumes that the values of measured variables in the population of interest are described by a probability distribution. If several variables are measured (say the length and weight of a cockroach), the population is described by a multivariate probability distribution. When the forces of good prevail, the fortunate statistician has data consisting of observed values of independent random variables, each having the population probability distribution. The statistical design of sampling and experimentation is intended to produce this happy state of affairs or some moderate complication of it. We will assume that, whether by design

* Preparation of this paper was supported by the Air Force Office of Scientific Research under Grant AFOSR-77-3291.

or (this is risky) by good fortune, the data collection process yields independent random variables $X_1, \ldots, X_n$ having a common probability distribution. This distribution is unknown—that's the distinction between statistics and probability theory. Let F denote the unknown distribution function (d.f.) of any single X_j.

It is clear that a classical statistical problem is "Which probability models adequately describe the data?" This question can be asked for descriptive purposes or as a preliminary to formal inference from the data. Particularly in the latter case, the statistician may have in mind a specific family of probability distributions (such as the normal family) and the more exact question "Do the data support or impugn the hypothesis that the population distribution is a member of this family?" Most common families of distributions have d.f.'s of specified functional form indexed by a (real or vector) parameter. For example, an individual member of the univariate normal family of distributions is specified by the values of the mean μ and standard deviation σ. If $G(\cdot|\theta)$ is a family of d.f.'s indexed by a parameter θ running over a parameter space Ω, we have now formulated the following problem.

Given independent random variables $X_1, \ldots, X_n$ having common unknown d.f. F, *test the hypothesis*

$$H_0 : F(\cdot) = G(\cdot|\theta) \qquad \textit{for some } \theta \textit{ in } \Omega.$$

This is the problem of *goodness of fit*. Notice that in practice the observations X_j will often be multivariate, and that the null hypothesis will usually be composite (that is, the family $G(\cdot|\theta)$ will contain more than one member). Notice also that although we have stated the problem in terms of hypothesis testing, it will rarely be sensible to simply accept or reject at the usual significance levels such as $\alpha = .05$. In particular, if we test fit to (say) the univariate normal family as a preliminary to a further analysis which assumes normality, we should surely not cling to the assumption of normality until the evidence against it is significant at the five percent level. Some applied statisticians favor using an α of .20 or .25 for such preliminary tests. The real difficulty is that the H_0 in the problem of fit does not have the status ("The statement we hope to find evidence against") ascribed to null hypotheses in standard tests of significance. Nonetheless, the attained significance level of a test of fit is at least a descriptive measure of the distance of the data from the hypothesized family of distributions. We will therefore study the theory of some tests of fit without further ventures into the wilderness of philosophies of inference.

The oldest family of tests of fit was fathered by Karl Pearson in 1900. During the preceding decade, Pearson had developed families of probability distributions in the course of his work on *Mathematical Contributions to the Theory of Evolution.* He now wished to see which of these fit his data, rather than simply assuming that all biological variables are normally distributed. Statistics as a discipline was in its infancy in 1900. Many results and methods which would form part of the new science were scattered through the work of Gauss, Laplace, Lagrange and others, but these results were not collected and unified, and were often unknown to statisticians such as Pearson. The binomial distributions and their approximation by normal distributions were well known; the chi-square distributions were known as the distributions of sums of squares of independent normal random variables; and the multivariate normal distributions had only recently become familiar. These last distributions will play a major role in our study. Pearson knew the p-variate normal distribution with mean vector $\boldsymbol{\mu}$ and nonsingular covariance matrix $\boldsymbol{\Sigma}$ as the distribution having density function of the form

$$f(\mathbf{y}') = ce^{-(1/2)(\mathbf{y}-\boldsymbol{\mu})'\boldsymbol{\Sigma}^{-1}(\mathbf{y}-\boldsymbol{\mu})}. \tag{1}$$

Here $\mathbf{y}' = (y_1, \ldots, y_p)$ is the p-variate argument of the density function. If $\mathbf{Y} = (Y_1, \ldots, Y_p)'$ is a random variable having this distribution, we will write $\mathbf{Y} \sim N_p(\boldsymbol{\mu}, \boldsymbol{\Sigma})$ to express this fact.

Pearson sought first to test the simple null hypothesis that univariate observations $X_1, \ldots, X_n$ have a given d.f. G. He partitioned the line into cells $E_1, \ldots, E_M$ and based his test on the frequencies $N_1, \ldots, N_M$ of observations falling in these cells. If the hypothesis is true and

$$p_i = P_G[X \text{ in } E_i] = \int_{E_i} dG(x),$$

then np_i is the expected frequency for E_i and the quantities $N_i - np_i$

measure the lack of fit between data and model. Pearson then argued:

(a) If $\mathbf{Y} \sim N_p(\mathbf{0}, \boldsymbol{\Sigma})$, then the quadratic form which appears in the exponent of the density function (1) has the same distribution as the sum of squares of p independent standard normal random variables. This is the chi-square distribution with p degrees of freedom, and we write $\mathbf{Y}'\boldsymbol{\Sigma}^{-1}\mathbf{Y} \sim \chi^2(p)$.

(b) By the DeMoivre-Laplace normal approximation to the binomial distributions, each $N_i - np_i$ is approximately normal when the number of observations n is large.

(c) Computing variances and covariances for $\mathbf{Y} = (N_1 - np_1, \ldots, N_{M-1} - np_{M-1})'$ and inverting the covariance matrix shows that

$$\mathbf{Y}'\boldsymbol{\Sigma}^{-1}\mathbf{Y} = \sum_{i=1}^{M} \frac{(N_i - np_i)^2}{np_i},$$

and this statistic therefore has approximately the $\chi^2(M-1)$ distribution when the null hypothesis is true and n is large. Large values of this statistic (i.e., values in the upper tail of the $\chi^2(M-1)$ distribution) are evidence of lack of fit.

This argument contains some minor mathematical gaps: it ignores the distinction between approximate normality of each $N_i - np_i$ and approximate multivariate normality of the vector $\mathbf{Y}$, and does not show how (a) implies that when $\mathbf{Y}$ is *approximately* $N_{M-1}(\mathbf{0}, \boldsymbol{\Sigma})$, then $\mathbf{Y}'\boldsymbol{\Sigma}^{-1}\mathbf{Y}$ is approximately $\chi^2(M-1)$. But the argument is in the best spirit of pre-Weierstrass mathematics, needing only a few technicalities to become rigorous. Pearson's proof shows, for example, why his famous chi-square statistic does not have the variance $np_i(1 - p_i)$ of N_i in the denominator of the ith summand. More important, the idea of reducing the general problem of fit to a combinatorial setting (counting numbers of observations in each of M cells) was of lasting significance. Chi-square tests remain among the most common tests of fit, largely because of the flexibility of Pearson's idea. If, for example, observations $\mathbf{X}_j$ and the cells E_i are multidimensional, the distribution of the cell frequencies N_i and the form and theory of the Pearson chi-square statistic are unchanged.

Now of course the null hypothesis in a problem of fit is generally that the d.f. of the observations falls in a family $\{G(\cdot|\boldsymbol{\theta}) : \boldsymbol{\theta} \text{ in } \Omega\}$ of d.f.'s. In this case, the cell probabilities depend on the unknown parameter $\boldsymbol{\theta}$,

$$p_i(\boldsymbol{\theta}) = \int_{E_i} dG(\mathbf{x}|\boldsymbol{\theta}).$$

Pearson proposed to estimate the unknown parameter from the data by some function $\boldsymbol{\theta}_n = \boldsymbol{\theta}_n(\mathbf{X}_1, \ldots, \mathbf{X}_n)$. In testing fit to the univariate normal family, for example, the parameter is $\boldsymbol{\theta} = (\mu, \sigma)$ and the population mean μ and standard deviation σ are commonly estimated by the mean and standard deviation of the sample. The Pearson statistic now becomes

$$\sum_{i=1}^{M} \frac{[N_i - np_i(\boldsymbol{\theta}_n)]^2}{np_i(\boldsymbol{\theta}_n)}. \tag{2}$$

That is, we test the fit of the data to the d.f. $G(\cdot|\boldsymbol{\theta}_n)$ having the estimated parameter value.

Unfortunately, as mathematicians learned before statisticians, pre-Weierstrass mathematics has its limitations. Even when the estimator $\boldsymbol{\theta}_n$ approaches the true value of $\boldsymbol{\theta}$ as the sample size n increases, the statistic (2) does *not* then have approximately the $\chi^2(M-1)$ distribution, as Pearson believed it did. Since statistical methods are actually used in the real world, observant users began to suspect that something was amiss. Some even did extensive simulations (quite a chore in those pre-computer days) to compare Pearson's theoretical distribution with the observed distribution of his statistic. It was not until 1924 that R. A. Fisher showed that the statistic (2) does not have approximately the $\chi^2(M-1)$ distribution in large samples, and that the distribution it does have depends on how the unknown parameter is estimated. If $\boldsymbol{\theta}_n$ is the value of $\boldsymbol{\theta}$ which minimizes the statistic (2) for given N_i (so that $G(\cdot|\boldsymbol{\theta}_n)$ is the closest d.f. in the hypothesized family to the data by this measure of distance), Fisher showed that the approximate distribution is $\chi^2(M-m-1)$ when $\boldsymbol{\theta}$ is m-dimensional. When other methods of estimation are used, this conclusion is false. It is false, for example, when the sample mean and standard deviation are used in testing fit to the univariate normal family. Only since the 1950s has a rigorous study of chi-square statistics with general estimators $\boldsymbol{\theta}_n$ been made, and solutions obtained to many practical and mathematical problems concerning these statistics.

Our study of the modern theory of chi-square tests of fit will touch on several other topics in statistics which are important in their own right. We must first acquaint ourselves with the multivariate normal family of distributions. And since chi-square tests are large-sample tests, based on the limiting multivariate normal distribution of the cell frequencies, some of the basic techniques of statistical large sample theory must be mentioned. Finally, Pearson's construction of the proper quadratic form in the cell frequencies is the genesis of some familiar statistical procedures other than tests of fit. In all of this, we hope not to entirely lose sight of the interplay between theory and practice which gives statistics its vitality.

2. THE MULTIVARIATE NORMAL DISTRIBUTIONS

The multivariate normal family plays a role in the study of multidimensional data analogous to that played by the univariate normal family in one dimension. These distributions are not only important probability models in their own right, but because of the central limit theorem serve as large sample approximations to other models. We will *not* define these distributions by the density function (1) for two reasons. First, that definition is awkward due to the complexity of the density function. More important, a distribution in Euclidean p-space R^p does not have a density function in R^p if it assigns probability one to a set of measure zero. Such *singular distributions*—other than the discrete distributions supported on a countable set—are somewhat pathological in one dimension, and play little role in probability modeling there. But in higher dimensions it is quite common for random variables to be dependent in such a way that with probability 1 their values fall in a lower-dimensional hyperplane and their joint distribution is thus singular. The cell frequencies $N_1, \ldots, N_M$ in a chi-square test, for example, satisfy $\sum_{i=1}^{M} N_i = n$ and so take values only in this $(M - 1)$-dimensional hyperplane. In Section 1 we followed Pearson in working with the nonsingular distribution of $(N_1, \ldots, N_{M-1})'$, but this mode of escape is more awkward in other settings.

To statisticians, the most useful definition of a probability distribution is often a *representational definition*—a statement of what random variable has the distribution. For example, the $\chi^2(p)$ distribution is that of $\sum_{i=1}^{p} Z_i^2$ where $Z_1, \ldots, Z_p$ are independent $N(0, 1)$ random variables. In this spirit, *the $N_p(\boldsymbol{\mu}, \boldsymbol{\Sigma})$ distribution is defined as the distribution of the random variable*

$$\mathbf{Y} = \mathbf{AZ} + \boldsymbol{\mu} \tag{3}$$

where $\mathbf{Z} = (Z_1, \ldots, Z_m)'$ and $Z_i \sim N(0, 1)$ and are independent, $\boldsymbol{\mu} = (\mu_1, \ldots, \mu_p)'$ is the vector of means, and $\mathbf{A}$ is any $p \times m$ matrix satisfying $\mathbf{AA}' = \boldsymbol{\Sigma}$. That is, the multivariate normal distributions are the distributions of affine transformations of a set of independent standard normal random variables.

It is easy to check that $\boldsymbol{\mu}$ and $\mathbf{AA}' = \boldsymbol{\Sigma}$ are in fact the mean vector and covariance matrix of the p-variate random vector defined in (3). To justify this definition of $N_p(\boldsymbol{\mu}, \boldsymbol{\Sigma})$, we must show that $\mathbf{Y}$ so defined has the *same* distribution for any m and any $p \times m$ matrix $\mathbf{A}$ satisfying $\mathbf{AA}' = \boldsymbol{\Sigma}$. To justify the notation $N_p(\boldsymbol{\mu}, \boldsymbol{\Sigma})$, we must show that this family is parameterized by $(\boldsymbol{\mu}, \boldsymbol{\Sigma})$ alone. Both of these facts follow from a computation of the characteristic function (Fourier transform) of the distribution of $\mathbf{Y}$. This computation also illustrates the convenience of the representational definition.

The characteristic function of $\mathbf{Y}$ is the function of p real variables $\mathbf{t}' = (t_1, \ldots, t_p)$ defined by

$$\begin{aligned} \varphi_{\mathbf{Y}}(\mathbf{t}') &= E[e^{i\mathbf{t}'\mathbf{Y}}] \\ &= E[e^{i(\mathbf{t}'\mathbf{AZ} + \mathbf{t}'\boldsymbol{\mu})}] \\ &= e^{i\mathbf{t}'\boldsymbol{\mu}} E[e^{i\mathbf{t}'\mathbf{AZ}}]. \end{aligned}$$

Now since the characteristic function of any $Z_j \sim N(0, 1)$ is easily computed to be

$$E[e^{isZ_j}] = e^{-(1/2)s^2} \qquad -\infty < s < \infty$$

and $Z_1, \ldots, Z_m$ are independent,

$$\begin{aligned}\varphi_{\mathbf{Y}}(\mathbf{t}') &= e^{i\mathbf{t}'\boldsymbol{\mu}} E[e^{i\{(\mathbf{t}'\mathbf{A})_1 Z_1 + \cdots + (\mathbf{t}'\mathbf{A})_m Z_m\}}] \\ &= e^{i\mathbf{t}'\boldsymbol{\mu}} \prod_{j=1}^{m} E[e^{i(\mathbf{t}'\mathbf{A})_j Z_j}] \\ &= e^{i\mathbf{t}'\boldsymbol{\mu}} \prod_{j=1}^{m} e^{-(1/2)[(\mathbf{t}'\mathbf{A})_j]^2} \\ &= e^{i\mathbf{t}'\boldsymbol{\mu} - (1/2)\mathbf{t}'\mathbf{A}\mathbf{A}'\mathbf{t}} = e^{i\mathbf{t}'\boldsymbol{\mu} - (1/2)\mathbf{t}'\boldsymbol{\Sigma}\mathbf{t}}.\end{aligned}$$

This characteristic function is the same for all m and $\mathbf{A}$ with $\mathbf{AA}' = \boldsymbol{\Sigma}$, and is parameterized by $\boldsymbol{\mu}$ and $\boldsymbol{\Sigma}$. Since the characteristic function uniquely determines a probability distribution, $N_p(\boldsymbol{\mu}, \boldsymbol{\Sigma})$ is well defined.

The definition (3) is geometrically transparent. The distribution of $\mathbf{Z}$ is nonsingular in R^m—indeed, has all of R^m as its support. If the linear transformation $\mathbf{A}: R^m \to R^p$ has full rank p, then the distribution $N_p(\boldsymbol{\mu}, \boldsymbol{\Sigma})$ of $\mathbf{Y}$ is nonsingular and has R^p as its support. If the rank of $\mathbf{A}$ is $r < p$, then $N_p(\boldsymbol{\mu}, \boldsymbol{\Sigma})$ is singular and is supported on the r-dimensional hyperplane in R^p obtained by translating the range of $\mathbf{A}$ by $\boldsymbol{\mu}$. Since the range of $\mathbf{AA}' = \boldsymbol{\Sigma}$ is the same as the range of $\mathbf{A}$, *$N_p(\boldsymbol{\mu}, \boldsymbol{\Sigma})$ is a nonsingular distribution if and only if the covariance matrix $\boldsymbol{\Sigma}$ is nonsingular. The support of $N_p(\boldsymbol{\mu}, \boldsymbol{\Sigma})$ is a hyperplane in R^p with dimension equal to the rank of $\boldsymbol{\Sigma}$.*

Many properties of the multivariate normal distributions follow easily from (3). For example, if $\mathbf{B}$ is any $s \times p$ matrix, then applying $\mathbf{B}$ to both sides of (3) shows at once that $\mathbf{BY} \sim N_s(\mathbf{B}\boldsymbol{\mu}, \mathbf{B}\boldsymbol{\Sigma}\mathbf{B}')$. That is, *any set of linear combinations of jointly normal random variables has again a multivariate normal joint distribution.* In particular, any single linear combination is univariate normal, and this includes each individual component Y_i of $\mathbf{Y}$. That the random variable $\mathbf{Y}$ has density function of the form (1) when $\boldsymbol{\Sigma}$ is nonsingular can be deduced by a change of variables in the known density function of $\mathbf{Z}$.

The joint distribution of a set of independent univariate normal random variables is a multivariate normal distribution with a diagonal covariance matrix $\boldsymbol{\Sigma}$. (Because $N(\mu_i, \sigma_i^2)$ is the distribution of $Y_i = \sigma_i Z_i + \mu_i$ for $Z_i \sim N(0, 1)$, so that (3) fits $\mathbf{Y} = (Y_1, \ldots, Y_p)'$.) Since $N_p(\boldsymbol{\mu}, \boldsymbol{\Sigma})$ is determined by $\boldsymbol{\mu}$ and $\boldsymbol{\Sigma}$, it follows that *random variables having a multivariate normal joint distribution are independent if and only if they are uncorrelated.* Independence in a multivariate normal setting can therefore be established by simply computing covariances. If $\mathbf{I}_p$ denotes the $p \times p$ identity matrix, $N_p(\mathbf{0}, \mathbf{I}_p)$ now denotes the distribution of a set of p independent $N(0, 1)$ random variables. If $\mathbf{Z} \sim N_p(\mathbf{0}, \mathbf{I}_p)$ and $\mathbf{P}$ is a $p \times p$ orthogonal matrix, it follows that $\mathbf{PZ} \sim N_p(\mathbf{0}, \mathbf{I}_p)$ once again.

For our study of chi-square tests, we are particularly interested in quadratic forms in multivariate normal random variables. The representational approach reduces this to the study of quadratic forms in independent $N(0, 1)$ random variables. We know one fact about such forms: if $\mathbf{Z} \sim N_p(\mathbf{0}, \mathbf{I}_p)$ then the sum of squares $\mathbf{Z}'\mathbf{Z} = \sum_{i=1}^p Z_i^2$ has the $\chi^2(p)$ distribution. Two potential generalizations of this fact come to mind at once. When $\mathbf{Y} \sim N_p(\mathbf{0}, \mathbf{\Sigma})$ we can ask (1) What quadratic forms $\mathbf{Y}'\mathbf{CY}$ have chi-square distributions? (2) What is the distribution of the sum of squares $\mathbf{Y}'\mathbf{Y}$? Since partial sums of squares in the $N_p(\mathbf{0}, \mathbf{I}_p)$ case have $\chi^2(k)$ distributions for $k < p$, the first question might be specialized to: What quadratic forms $\mathbf{Y}'\mathbf{CY}$ have the $\chi^2(r)$ distribution for r as large as possible?

It is convenient for the study of quadratic forms to use a particular representation of $\mathbf{Y} \sim N_p(\mathbf{0}, \mathbf{\Sigma})$ based on the fact that *any nonnegative definite symmetric matrix (such as an arbitrary covariance matrix* $\mathbf{\Sigma}$*) has a unique nonnegative definite symmetric square root.* To see this, note first that any square root $\mathbf{\Sigma}^{1/2}$ commutes with its square $\mathbf{\Sigma}$, so that if $\mathbf{\Sigma}^{1/2}$ is symmetric, $\mathbf{\Sigma}^{1/2}$ and $\mathbf{\Sigma}$ can be simultaneously diagonalized by some orthogonal matrix $\mathbf{P}$. From

$$\mathbf{P\Sigma P}' = \begin{pmatrix} \sigma_1 & & \\ & \ddots & \\ & & \sigma_p \end{pmatrix} = \mathbf{D} \tag{4}$$

$$\mathbf{P\Sigma}^{1/2}\mathbf{P}' = \begin{pmatrix} \nu_1 & & \\ & \ddots & \\ & & \nu_p \end{pmatrix}$$

and $(\mathbf{\Sigma}^{1/2})^2 = \mathbf{\Sigma}$ it follows that $\nu_i = \sigma_i^{1/2}$ and hence that all symmetric nonnegative definite square roots have the form

$$\mathbf{\Sigma}^{1/2} = \mathbf{P}'\begin{pmatrix} \sigma_1^{1/2} & & \\ & \ddots & \\ & & \sigma_p^{1/2} \end{pmatrix}\mathbf{P}$$

for $\mathbf{P}$ and $\mathbf{D}$ satisfying (4) and nonnegative square roots $\sigma_i^{1/2}$. This also shows that such a $\mathbf{\Sigma}^{1/2}$ exists. The $\mathbf{P}$ and $\mathbf{D}$ in (4) are not unique, but are determined up to permutations of the σ_i and choice of corresponding orthogonal characteristic vectors as rows of $\mathbf{P}$. Since $\mathbf{\Sigma}^{1/2}$ is unchanged by these choices, it is unique.

We can thus represent $\mathbf{Y} \sim N_p(\mathbf{0}, \mathbf{\Sigma})$ as $\mathbf{Y} = \mathbf{\Sigma}^{1/2}\mathbf{Z}$. The distribution of quadratic forms in $\mathbf{Y}$ is completely described by the following result.

THEOREM 1: *Suppose that* $\mathbf{Y} \sim N_p(\mathbf{0}, \mathbf{\Sigma})$ *and that* $\mathbf{C}$ *is any* $p \times p$ *symmetric matrix. Then the quadratic form* $\mathbf{Y}'\mathbf{C}\mathbf{Y}$ *has the distribution of* $\sum_{i=1}^{p} \lambda_i Z_i^2$, *where the* Z_i *are independent* $N(0, 1)$ *random variables and the* λ_i *are the characteristic roots of* $\mathbf{\Sigma}^{1/2}\mathbf{C}\mathbf{\Sigma}^{1/2}$.

Proof: Since $\mathbf{Y} = \mathbf{\Sigma}^{1/2}\mathbf{Z}$,

$$\mathbf{Y}'\mathbf{C}\mathbf{Y} = \mathbf{Z}'\mathbf{\Sigma}^{1/2}\mathbf{C}\mathbf{\Sigma}^{1/2}\mathbf{Z} = \mathbf{Z}'\mathbf{Q}\mathbf{Z}.$$

Because $\mathbf{Q} = \mathbf{\Sigma}^{1/2}\mathbf{C}\mathbf{\Sigma}^{1/2}$ is symmetric, there is an orthogonal matrix $\mathbf{P}$ such that $\mathbf{P}\mathbf{Q}\mathbf{P}' = \mathbf{D}$, where $\mathbf{D}$ is diagonal with the λ_i as diagonal elements. So $\mathbf{Q} = \mathbf{P}'\mathbf{D}\mathbf{P}$ and

$$\mathbf{Y}'\mathbf{C}\mathbf{Y} = \mathbf{Z}'\mathbf{Q}\mathbf{Z} = (\mathbf{P}\mathbf{Z})'\mathbf{D}(\mathbf{P}\mathbf{Z}).$$

The right side above is $\sum_{i=1}^{p} \lambda_i (Z_i^*)^2$ where Z_i^* is the ith component of $\mathbf{P}\mathbf{Z}$. Since $\mathbf{P}\mathbf{Z} \sim N_p(\mathbf{0}, \mathbf{I}_p)$, this is a representation of the desired form.

When $\mathbf{X} \sim N_p(\boldsymbol{\mu}, \mathbf{\Sigma})$ and $\mathbf{\Sigma}$ is nonsingular, $\mathbf{Y} = \mathbf{X} - \boldsymbol{\mu} \sim N_p(\mathbf{0}, \mathbf{\Sigma})$ and we obtain as a corollary of Theorem 1 Pearson's result that the quadratic form $(\mathbf{X} - \boldsymbol{\mu})'\mathbf{\Sigma}^{-1}(\mathbf{X} - \boldsymbol{\mu})$ appearing in the exponent of the density function has the distribution of $\sum_{i=1}^{p} Z_i^2$, which is $\chi^2(p)$. This answers our first question when $\mathbf{\Sigma}$ is nonsingular. The answer to the second question, concerning the distribution of the sum of squares $\mathbf{Y}'\mathbf{Y}$, is answered by setting $\mathbf{C} = \mathbf{I}_p$ in Theorem 1. The distribution is that of $\sum_{i=1}^{p} \lambda_i Z_i^2$, where the λ_i are now the characteristic roots of $\mathbf{\Sigma}$ itself.

We have yet to answer the first question fully by extending Pearson's recipe to the singular case. Since the rank of $\mathbf{\Sigma}^{1/2}\mathbf{C}\mathbf{\Sigma}^{1/2}$ cannot exceed the rank (say r) of $\mathbf{\Sigma}$ and $\mathbf{\Sigma}^{1/2}$, it is clear that if $\mathbf{Y}'\mathbf{C}\mathbf{Y} \sim \chi^2(k)$, then

$k \leqslant r$. Theorem 1 implies that $\mathbf{Y}'\mathbf{C}\mathbf{Y} \sim \chi^2(k)$ if and only if $\mathbf{\Sigma}^{1/2}\mathbf{C}\mathbf{\Sigma}^{1/2}$ is idempotent of rank k, and a little matrix manipulation shows that a sufficient condition for this is that $\mathbf{\Sigma C}$ be idempotent of rank k. This general result is not very helpful in the search for $\mathbf{C}$ such that $\mathbf{Y}'\mathbf{C}\mathbf{Y} \sim \chi^2(r)$. Pearson's result involved the inverse $\mathbf{\Sigma}^{-1}$. It turns out that an "inverse" for singular matrices neatly generalizes his recipe.

For an arbitrary $n \times m$ real matrix $\mathbf{A}$, a *generalized inverse* of $\mathbf{A}$ can be defined by any of the following statements.

(a) A generalized inverse of $\mathbf{A}$ is any $m \times n$ matrix $\mathbf{G}$ such that $\mathbf{x} = \mathbf{G}\mathbf{y}$ solves the equations $\mathbf{y} = \mathbf{A}\mathbf{x}$ for any $\mathbf{y}$ in the range (column space) of $\mathbf{A}$.

(b) A generalized inverse of $\mathbf{A}$ is any $m \times n$ matrix $\mathbf{G}$ satisfying $\mathbf{AGA} = \mathbf{A}$.

(c) A generalized inverse of $\mathbf{A}$ is any $m \times n$ matrix $\mathbf{G}$ such that $\mathbf{AG}$ is a projection onto the range of $\mathbf{A}$.

It is not hard to show that these definitions are equivalent. Definition (a) justifies the concept in terms of solving consistent sets of linear equations with matrix $\mathbf{A}$. Definition (b) is convenient for matrix manipulation, while (c) gives some geometric insight. Note that in (c) "projection" does not mean the unique orthogonal projection onto the range of $\mathbf{A}$, but any idempotent matrix with this range. When $\mathbf{A}$ is not a nonsingular square matrix, it possesses many generalized inverses. Any generalized inverse of $\mathbf{A}$ will be denoted by $\mathbf{A}^-$. Generalized inverses are widely used to provide a unified notation for linear statistical problems when matrices may be singular. The following theorem and its proof illustrate the convenience of this notion.

THEOREM 2: *Suppose that* $\mathbf{Y} \sim N_p(\mathbf{0}, \mathbf{\Sigma})$ *and* $\mathbf{\Sigma}$ *has rank* r. *Then*

(a) *With probability* 1, $\mathbf{Y}'\mathbf{\Sigma}^-\mathbf{Y}$ *is the same for all choices of* $\mathbf{\Sigma}^-$.

(b) $\mathbf{Y}'\mathbf{\Sigma}^-\mathbf{Y} \sim \chi^2(r)$ *and is the unique quadratic form having this distribution.*

Proof: (a) If $\mathbf{x}$ is any vector in the range of $\mathbf{\Sigma}$, so that $\mathbf{x} = \mathbf{\Sigma y}$ for some $\mathbf{y}$, then

$$\mathbf{x}'\mathbf{\Sigma}^-\mathbf{x} = \mathbf{y}'\mathbf{\Sigma}\mathbf{\Sigma}^-\mathbf{\Sigma}\mathbf{y} = \mathbf{y}'\mathbf{\Sigma}\mathbf{y}$$

by definition (b) and symmetry of $\mathbf{\Sigma}$. So $\mathbf{x}'\mathbf{\Sigma}^-\mathbf{x}$ is the same number for all choices of $\mathbf{\Sigma}^-$ whenever $\mathbf{x}$ is in the range of $\mathbf{\Sigma}$. But $\mathbf{Y} = \mathbf{\Sigma}^{1/2}\mathbf{Z}$ is in the range of $\mathbf{\Sigma}$ (which is the same as that of $\mathbf{\Sigma}^{1/2}$) with probability 1.

(b) Since $\mathbf{Y}'\mathbf{\Sigma}^-\mathbf{Y}$ is the same for all choices of $\mathbf{\Sigma}^-$, we can choose a convenient generalized inverse. If $\mathbf{\Sigma}$ has rank r and positive characteristic roots $d_1, \ldots, d_r$, there is an orthogonal $\mathbf{P}$ such that

$$\mathbf{P\Sigma P}' = \begin{pmatrix} d_1 & & & & & \\ & \ddots & & & & \\ & & d_r & & & \\ & & & 0 & & \\ & & & & \ddots & \\ & & & & & 0 \end{pmatrix}.$$

An obvious choice of generalized inverse is

$$\mathbf{\Sigma}^- = \mathbf{P}'\begin{pmatrix} d_1^{-1} & & & & & \\ & \ddots & & & & \\ & & d_r^{-1} & & & \\ & & & 0 & & \\ & & & & \ddots & \\ & & & & & 0 \end{pmatrix}\mathbf{P}.$$

Since

$$\mathbf{\Sigma}^{1/2} = \mathbf{P}'\begin{pmatrix} d_1^{1/2} & & & & & \\ & \ddots & & & & \\ & & d_r^{1/2} & & & \\ & & & 0 & & \\ & & & & \ddots & \\ & & & & & 0 \end{pmatrix}\mathbf{P}$$

we obtain

$$\mathbf{Y}'\mathbf{\Sigma}^-\mathbf{Y} = \mathbf{Z}'\mathbf{\Sigma}^{1/2}\mathbf{\Sigma}^-\mathbf{\Sigma}^{1/2}\mathbf{Z} = \sum_{i=1}^{r} (Z_i^*)^2$$

where $\mathbf{Z}^* = \mathbf{PZ} \sim N_p(\mathbf{0}, \mathbf{I}_p)$. Thus $\mathbf{Y}'\mathbf{\Sigma}^-\mathbf{Y} \sim \chi^2(r)$.

It remains to show that $\mathbf{Y}'\mathbf{CY} \sim \chi^2(r)$ implies that $\mathbf{C}$ is a generalized inverse of $\mathbf{\Sigma}$. By Theorem 1, $\mathbf{Y}'\mathbf{CY} \sim \chi^2(r)$ if and only if $\mathbf{\Sigma}^{1/2}\mathbf{C\Sigma}^{1/2}$ is idempotent of rank r. But then $\mathbf{\Sigma}^{1/2}\mathbf{C\Sigma}^{1/2}$ is a projection, and since its range is contained in the range of $\mathbf{\Sigma}^{1/2}$ and has the same dimension r, it is a projection onto the range of $\mathbf{\Sigma}^{1/2}$. A projection acts as the identity transformation on its range, so

$$\mathbf{\Sigma C \Sigma} = \mathbf{\Sigma}^{1/2}(\mathbf{\Sigma}^{1/2}\mathbf{C}\mathbf{\Sigma}^{1/2})\mathbf{\Sigma}^{1/2} = \mathbf{\Sigma}^{1/2}\mathbf{\Sigma}^{1/2} = \mathbf{\Sigma}$$

and $\mathbf{C}$ satisfies definition (b) of $\mathbf{\Sigma}^{-}$.

When $\mathbf{Y} \sim N_p(\boldsymbol{\mu}, \mathbf{\Sigma})$, the representation $\mathbf{Y} = \mathbf{\Sigma}^{1/2}\mathbf{Z} + \boldsymbol{\mu}$ can be applied to the study of quadratic forms in $\mathbf{Y}$. Repeating the argument of Theorem 1 shows that $\mathbf{Y}'\mathbf{C}\mathbf{Y}$ has the distribution of a random variable of the form

$$\sum_{i=1}^{p} \lambda_i Z_i^2 + 2 \sum_{i=1}^{p} b_i Z_i + c \tag{5}$$

where $\mathbf{Z} \sim N_p(\mathbf{0}, \mathbf{I}_p)$ and the λ_i are as in Theorem 1. These distributions have no neat classification. We will make only one foray into this "noncentral case," to look again at Pearson's recipe.

When $\mathbf{Z} \sim N_p(\mathbf{0}, \mathbf{I}_p)$, $\mathbf{Z}'\mathbf{Z} \sim \chi^2(p)$ by definition. When $\mathbf{Y} \sim N_p(\boldsymbol{\mu}, \mathbf{I}_p)$, the distribution of $\mathbf{Y}'\mathbf{Y}$, or equivalently of $(\mathbf{Z} + \boldsymbol{\mu})'(\mathbf{Z} + \boldsymbol{\mu})$, is defined to be the *noncentral chi-square distribution with p degrees of freedom and noncentrality parameter* $\delta = \boldsymbol{\mu}'\boldsymbol{\mu}$. (Since the statistic is the square of the distance of $\mathbf{Z}$ from the point $-\boldsymbol{\mu}$ in R^p, it follows from the circular symmetry of the density function of $\mathbf{Z}$ that this distribution depends only on the distance of $-\boldsymbol{\mu}$ from the origin. Thus parameterizing the distribution by (p, δ) is justified). We will use the notation $\mathbf{Y}'\mathbf{Y} \sim \chi^2(p, \delta)$.

Suppose now that $\mathbf{Y} \sim N_p(\boldsymbol{\mu}, \mathbf{\Sigma})$ and $\mathbf{\Sigma}$ has rank r. What then is the distribution of the generalized Pearson statistic $\mathbf{Y}'\mathbf{\Sigma}^{-}\mathbf{Y}$? Alas, since $\mathbf{Y} = \mathbf{\Sigma}^{1/2}\mathbf{Z} + \boldsymbol{\mu}$, $\mathbf{Y}$ is not in the range of $\mathbf{\Sigma}$ unless $\boldsymbol{\mu}$ is, so that the quadratic form $\mathbf{Y}'\mathbf{\Sigma}^{-}\mathbf{Y}$ changes with the choice of $\mathbf{\Sigma}^{-}$. If $\boldsymbol{\mu}$ is in the range of $\mathbf{\Sigma}$, we can write $\boldsymbol{\mu} = \mathbf{\Sigma}^{1/2}\boldsymbol{\nu}$ and follow the argument of Theorem 2 to show that $\mathbf{Y}'\mathbf{\Sigma}^{-}\mathbf{Y}$ is well defined and that

$$\begin{aligned}\mathbf{Y}'\mathbf{\Sigma}^{-}\mathbf{Y} &= (\mathbf{Z} + \boldsymbol{\nu})'\mathbf{\Sigma}^{1/2}\mathbf{\Sigma}^{-}\mathbf{\Sigma}^{1/2}(\mathbf{Z} + \boldsymbol{\nu}) \\ &= \sum_{i=1}^{r} (Z_i + \nu_i)^2 \sim \chi^2(r, \delta)\end{aligned}$$

where $\delta = \sum_{i=1}^{r} \nu_i^2$. But by the same argument,

$$\sum_{i=1}^{r} \nu_i^2 = \boldsymbol{\nu}'\mathbf{\Sigma}^{1/2}\mathbf{\Sigma}^{-}\mathbf{\Sigma}^{1/2}\boldsymbol{\nu} = \boldsymbol{\mu}'\mathbf{\Sigma}^{-}\boldsymbol{\mu}.$$

Thus $\mathbf{Y}'\mathbf{\Sigma}^-\mathbf{Y} \sim \chi^2(r, \boldsymbol{\mu}'\mathbf{\Sigma}^-\boldsymbol{\mu})$. When $\boldsymbol{\mu}$ is not in the range of $\mathbf{\Sigma}$, both the form and the distribution of $\mathbf{Y}'\mathbf{\Sigma}^-\mathbf{Y}$ vary with the choice of $\mathbf{\Sigma}^-$. Of course, when $\mathbf{\Sigma}$ is nonsingular these complications do not arise, and $\mathbf{Y}'\mathbf{\Sigma}^{-1}\mathbf{Y} \sim \chi^2(p, \boldsymbol{\mu}'\mathbf{\Sigma}^{-1}\boldsymbol{\mu})$ for any mean vector $\boldsymbol{\mu}$.

We have concentrated on cases in which the distribution of $\mathbf{Y}'\mathbf{C}\mathbf{Y}$ given by Theorem 1 (or more generally by (5)) reduces to a chi-square distribution. There are sound practical reasons for doing so, even though machine computation makes it feasible to produce tables of critical points for the distributions of $\sum_{i=1}^{p} \lambda_i Z_i^2$. Tests of fit based on quadratic forms in (approximately) multivariate normal random variables are the natural generalization of Pearson's chi-square test. These tests must compete for the attention of practical statisticians against special-purpose tests for fit to specific common families, and against general tests of fit based on the empirical distribution function (EDF tests). These competitors are usually more powerful than chi-square tests, but are also less flexible in adapting to unknown parameters and discrete or multivariate data. In particular, they require separate computation of critical points for each hypothesized family. (I will not mention—this is a rhetorical device I learned from Cicero—that the EDF tests break down almost completely when faced with hypothesized distributions which are multi-dimensional or are not location-scale families.) If a test of chi-square type also requires a special computation of critical points to be applicable to a given problem, we would usually be wiser to allot our computer time to an EDF test instead. Thus generalizations of Pearson's statistic lose much of their attractiveness if their critical points cannot be found in standard tables. In the light of Theorem 1, the relevant tables will be those for the chi-square distributions.

3. LARGE SAMPLE THEORY

Since the earliest days of statistics it has been noticed that complicated distributions often have simple approximations for large samples. The distribution of Pearson's chi-square statistic is an example. The use of chi-square tests, both as tests of fit and for other common applications, is based on approximating multinomial distributions by the multivariate normal distributions which are their limits as the sample size increases. We will therefore review some facts about statistical large sample theory. There are three major aspects to this theory. The first simply asks questions of convergence: "What happens in the limit?" The second studies the approach to the limit by providing rates of convergence, asymptotic expansions, etc. The third considers the usefulness of the asymptotic forms provided by the first two parts of the subject as approximations to the fixed sample size truth. Explicit numerical calculation and simulation play large roles here. Only the first aspect of large sample theory will concern us, both for simplicity's sake and because (to make an appalling generalization) in the field of chi-square tests the second aspect has had little practical impact and the third has shown that use of limiting distributions is an adequate approximation for quite moderate sample sizes.

The most useful mode of convergence for statistical use is convergence in distribution. If $\mathbf{X}_1, \mathbf{X}_2, \ldots,$ are R^p-valued random variables, $\mathbf{X}_n$ having d.f. F_n, we say that the sequence *converges in distribution* to the distribution having d.f. F if $F_n(\mathbf{x}) \to F(\mathbf{x})$ for every continuity point $\mathbf{x}$ of F. Abusing notation to also denote by F, F_n the probability measures on R^p generated by these d.f.'s, convergence in distribution is equivalent to

$$\lim_n P[\mathbf{X}_n \text{ in } A] = \lim_n F_n(A) = F(A)$$

for all Borel sets A in R^p whose boundaries have probability zero under F. Thus $P[\mathbf{X}_n \text{ in } A]$ can be approximated by $F(A)$ for large n. Convergence in distribution to the distribution placing probability 1 on a single point $\mathbf{c}$ is equivalent to *convergence in probability* of $\mathbf{X}_n$ to $\mathbf{c}$. That is, for any $\epsilon > 0$, $P[|\mathbf{X}_n - \mathbf{c}| > \epsilon] \to 0$ as $n \to \infty$. (We write this $\mathbf{X}_n \to \mathbf{c}(P)$.) All of this is of course a province of the measure theory which underlies statistical theory and sometimes invades the conscious thought of the working statistician. A nice exposition in effortless generality appears in the first chapter of [1]. We require only one specific and two general facts about convergence in distribution. The specific fact is *the multivariate central limit theorem: If* $\mathbf{X}_1, \mathbf{X}_2, \ldots,$ *are independent* R^p*-valued random variables* *having a common distribution with vector of means* $\boldsymbol{\mu}$ *and finite* $p \times p$ *covariance matrix* $\boldsymbol{\Sigma}$, *and if* $\bar{\mathbf{X}}_n = n^{-1} \sum_1^n \mathbf{X}_i$, *then* $n^{1/2}(\bar{\mathbf{X}}_n - \boldsymbol{\mu})$ *converges in distribution to* $N_p(\mathbf{0}, \boldsymbol{\Sigma})$. This is written

$$n^{1/2}(\bar{\mathbf{X}}_n - \boldsymbol{\mu}) \xrightarrow{\mathscr{D}} N_p(\mathbf{0}, \boldsymbol{\Sigma}).$$

We often abuse notation and write instead

$$n^{1/2}(\bar{\mathbf{X}}_n - \boldsymbol{\mu}) \xrightarrow{\mathscr{D}} \mathbf{Y}$$

where $\mathbf{Y} \sim N_p(\mathbf{0}, \boldsymbol{\Sigma})$, even though convergence in distribution makes no statement about convergence of values of $n^{1/2}(\bar{\mathbf{X}}_n - \boldsymbol{\mu})$ or about any limiting random variable.

The essential general fact is *the continuity theorem: If* $\mathbf{Y}_n \xrightarrow{\mathscr{D}} \mathbf{Y}$ *and* $h : R^p \to R^k$ *is continuous with probability* 1 *with respect to the distribution of* $\mathbf{Y}$, *then* $h(\mathbf{Y}_n) \xrightarrow{\mathscr{D}} h(\mathbf{Y})$. The continuity theorem licenses our natural desire to conclude that when (say) $\mathbf{Y}_n$ is approximately $N_p(\mathbf{0}, \boldsymbol{\Sigma})$, then $\mathbf{Y}_n'\mathbf{C}\mathbf{Y}_n$ has approximately the distribution specified by Theorem 1. The central limit theorem provides us with a large supply of random variables which are approximately multivariate normal. The two together suffice to make rigorous Pearson's proof outlined in Section 1 above.

The second general fact is needed to still the clamoring voices of the pedants. *If* $\mathbf{X}_n \xrightarrow{\mathscr{D}} \mathbf{X}$ *and* $\mathbf{Y}_n \to \mathbf{c}(P)$, *then* $(\mathbf{X}_n, \mathbf{Y}_n) \xrightarrow{\mathscr{D}} (\mathbf{X}, \mathbf{c})$. That is, convergence of both marginal distributions of $(\mathbf{X}_n, \mathbf{Y}_n)$ suffices for convergence of the joint distribution *if* one sequence of marginal distributions has a degenerate limit. Convergence of marginal distributions in general gives no information about the joint distribution. The natural manipulations we wish to make are all licensed by these two general facts. For example, if $\mathbf{X}_n \xrightarrow{\mathscr{D}} \mathbf{X}$ and $\mathbf{R}_n \to \mathbf{0}(P)$, then $\mathbf{X}_n + \mathbf{R}_n \xrightarrow{\mathscr{D}} \mathbf{X}$, as reason and justice demand. For $(\mathbf{X}_n, \mathbf{R}_n) \xrightarrow{\mathscr{D}} (\mathbf{X}, \mathbf{0})$ by the second fact, and the continuity theorem now applies with $h(\mathbf{x}, \mathbf{y}) = \mathbf{x} + \mathbf{y}$.

The following section will provide examples in plenty of the way in which the three facts mentioned here combine with the law of large numbers and Taylor's theorem to form the elementary arithmetic of statistical large sample theory.

4. CHI-SQUARE TESTS OF FIT

Returning at last to the problem and notation of Section 1, we wish to test whether independent random variables $\mathbf{X}_1, \ldots, \mathbf{X}_n$ taking values in Euclidean p-space R^p have d.f. $G(\cdot|\boldsymbol{\theta})$ for some $\boldsymbol{\theta}$ in Ω, an open set in R^m. Partitioning R^p into M cells $E_1, \ldots, E_M$, we denote by N_i the number of $\mathbf{X}_1, \ldots, \mathbf{X}_n$ falling in E_i and by $p_i(\boldsymbol{\theta})$ the probability that a random variable with d.f. $G(\cdot|\boldsymbol{\theta})$ falls in E_i. The vector of standardized cell frequencies is the M-vector $\mathbf{V}_n(\boldsymbol{\theta})$ with ith component

$$\frac{N_i - np_i(\boldsymbol{\theta})}{[np_i(\boldsymbol{\theta})]^{1/2}}.$$

Finally, $\boldsymbol{\theta}$ is estimated from $\mathbf{X}_1, \ldots, \mathbf{X}_n$ by $\boldsymbol{\theta}_n = \boldsymbol{\theta}_n(\mathbf{X}_1, \ldots, \mathbf{X}_n)$, and $\mathbf{C}_n = \mathbf{C}_n(\mathbf{X}_1, \ldots, \mathbf{X}_n)$ is a possibly data-dependent nonnegative definite symmetric $M \times M$ matrix. *Statistics of chi-square type are statistics of the form*

$$\mathbf{V}_n(\boldsymbol{\theta}_n)'\mathbf{C}_n\mathbf{V}_n(\boldsymbol{\theta}_n), \tag{6}$$

that is, nonnegative definite quadratic forms in the standardized cell frequencies.

If the vector $\mathbf{V}_n(\boldsymbol{\theta}_n)$ has a limiting $N_M(\mathbf{0}, \boldsymbol{\Sigma}(\boldsymbol{\theta}_0))$ distribution when $G(\cdot|\boldsymbol{\theta}_0)$ is the true d.f., and if $\mathbf{C}_n \to \mathbf{C}(\boldsymbol{\theta}_0)(P)$, then the continuity theorem tells us that the limiting distributions of statistics of chi-square type under the null hypothesis are completely described by Theorem 1. Establishing asymptotic normality of $\mathbf{V}_n(\boldsymbol{\theta}_n)$ is therefore the primary mathematical hurdle in the theory of chi-square statistics. When $\boldsymbol{\theta}$, or more precisely the vector $\mathbf{p}(\boldsymbol{\theta}) = (p_1(\boldsymbol{\theta}), \ldots, p_M(\boldsymbol{\theta}))'$, is known, this hurdle is low indeed. For the N_i have a multinomial distribution with parameters n and $\mathbf{p}(\boldsymbol{\theta})$. The vector $(N_1, \ldots, N_M)$ can be expressed as the sum of n independent M-dimensional indicator variables $\boldsymbol{\delta}_1, \ldots, \boldsymbol{\delta}_n$ where $\boldsymbol{\delta}_j$ has ith component 1 and all others 0 when $\mathbf{X}_j$ falls in cell E_i. It follows from a computation of covariances and the multivariate central limit theorem that under $G(\cdot|\boldsymbol{\theta})$

$$\mathbf{V}_n(\boldsymbol{\theta}) \xrightarrow{\mathscr{D}} N_M(\mathbf{0}, \mathbf{I}_M - \mathbf{q}(\boldsymbol{\theta})\mathbf{q}(\boldsymbol{\theta})'), \tag{7}$$

where $\mathbf{I}_M$ is the $M \times M$ identity matrix and

$$\mathbf{q}(\boldsymbol{\theta}) = (p_1(\boldsymbol{\theta})^{1/2}, \ldots, p_M(\boldsymbol{\theta})^{1/2})'.$$

This is just the multivariate normal approximation to a multinomial distribution, expressed in a notation which will prove convenient for easy extension to the more common case when $\boldsymbol{\theta}$ must be estimated.

In that latter case, the asymptotic behavior of $\mathbf{V}_n(\boldsymbol{\theta}_n)$ will depend on that of $\boldsymbol{\theta}_n$, as Fisher recognized. Thus the large sample theory of chi-square statistics draws on the large sample theory of estimators, a main current of statistical theory since Fisher's time. Because of the importance of this subject, and to illustrate the application of the principles stated in Section 3, there follows an account of the large sample behavior of $\mathbf{V}_n(\boldsymbol{\theta}_n)$ and also of the *minimum chi-square estimator* $\bar{\boldsymbol{\theta}}_n$ used in the classical Pearson-Fisher test. Readers uninterested in this analytical detail may note only that $\bar{\boldsymbol{\theta}}_n$ behaves as described in (9) below, and then proceed to (14), which displays explicitly the large sample relation between $\mathbf{V}_n(\boldsymbol{\theta}_n)$ and $\boldsymbol{\theta}_n$.

For given $N_1, \ldots, N_M$ the minimum chi-square estimator is any value of $\boldsymbol{\theta}$ which minimizes the Pearson statistic

$$P_n(\boldsymbol{\theta}) = \mathbf{V}_n(\boldsymbol{\theta})'\mathbf{V}_n(\boldsymbol{\theta}) = \sum_{i=1}^{M} \frac{[N_i - np_i(\boldsymbol{\theta})]^2}{np_i(\boldsymbol{\theta})}.$$

It is intuitively clear, and not hard to prove, that in large samples $\bar{\boldsymbol{\theta}}_n$ is equivalent to the *minimum modified chi-square estimator* which minimizes the modified chi-square statistic

$$Q_n(\boldsymbol{\theta}) = \sum_{i=1}^{M} \frac{[N_i - np_i(\boldsymbol{\theta})]^2}{N_i}.$$

This estimator, which we assume to exist and be a measurable function of $N_1, \ldots, N_M$, will also be denoted by $\bar{\boldsymbol{\theta}}_n$. The use of Q_n rather than P_n saves us some inessential detail in the development below. The first question concerns the *consistency* of this estimator—does $\bar{\boldsymbol{\theta}}_n$ approach the true value of $\boldsymbol{\theta}$ as n increases?

LEMMA 1: *Suppose that $M \geqslant m$, that each $\partial p_i/\partial \theta_k$ is continuous at $\boldsymbol{\theta} = \boldsymbol{\theta}_0$, and that the $M \times m$ matrix $(\partial p_i/\partial \theta_k)(\boldsymbol{\theta}_0)$ has rank m.*

Suppose also that the function $\boldsymbol{\theta} \to \mathbf{p}(\boldsymbol{\theta})$ *from* R^m *to* R^M *has a continuous inverse at* $\boldsymbol{\theta}_0$. *Then any minimum modified chi-square estimator* $\bar{\boldsymbol{\theta}}_n$ *satisfies* $\bar{\boldsymbol{\theta}}_n \to \boldsymbol{\theta}_0(P)$ *when* $G(\cdot|\boldsymbol{\theta}_0)$ *is the true* d.f. *of the* $\mathbf{X}_j$.

Proof: By the law of large numbers,

$$N_i/n \to p_i(\boldsymbol{\theta}_0)(P) \qquad i = 1, \ldots, M \tag{8}$$

and therefore by the continuity theorem

$$Q_n(\boldsymbol{\theta}_0)/n = \sum_{i=1}^{M} \frac{[N_i/n - p_i(\boldsymbol{\theta}_0)]^2}{N_i/n} \to 0(P).$$

But by the definition of $\bar{\boldsymbol{\theta}}_n$,

$$0 \leqslant Q_n(\bar{\boldsymbol{\theta}}_n)/n \leqslant Q_n(\boldsymbol{\theta}_0)/n$$

and so $Q_n(\bar{\boldsymbol{\theta}}_n)/n \to 0(P)$. This can only happen if $N_i/n - p_i(\bar{\boldsymbol{\theta}}_n) \to 0(P)$ for $i = 1, \ldots, M$. This with (8) implies that $p(\bar{\boldsymbol{\theta}}_n) \to p(\boldsymbol{\theta}_0)(P)$, which in turn implies that $\bar{\boldsymbol{\theta}}_n \to \boldsymbol{\theta}_0(P)$ if the function $\boldsymbol{\theta} \to \mathbf{p}(\boldsymbol{\theta})$ from R^m to R^M has a continuous inverse at $\boldsymbol{\theta} = \boldsymbol{\theta}_0$, using the continuity theorem once again.

The actual large sample form of $\bar{\boldsymbol{\theta}}_n$ is given by the following theorem. The result is Fisher's, but a rigorous proof first appeared in Cramér's classic book [3] in 1946. The proof provides as a bonus an expression for $\mathbf{V}_n(\boldsymbol{\theta}_n)$ for general estimators $\boldsymbol{\theta}_n$. Denote by $\mathbf{B}(\boldsymbol{\theta})$ the $M \times m$ matrix with (i, k)th entry

$$p_i(\boldsymbol{\theta})^{-1/2} \frac{\partial p_i(\boldsymbol{\theta})}{\partial \theta_k}.$$

In analogy with the common $o(1)$ notation from analysis, $o_p(1)$ denotes any quantity converging in probability to zero as n increases. From this point we shall for brevity omit the argument $\boldsymbol{\theta}$ when $\boldsymbol{\theta} = \boldsymbol{\theta}_0$. Thus, for example, $\mathbf{B} = \mathbf{B}(\boldsymbol{\theta}_0)$ and $\mathbf{V}_n = \mathbf{V}_n(\boldsymbol{\theta}_0)$ in the statement of the following theorem.

THEOREM 3: *Under the conditions of Lemma* 1, *when* $G(\cdot|\boldsymbol{\theta}_0)$ *holds*,

$$n^{1/2}(\bar{\boldsymbol{\theta}}_n - \boldsymbol{\theta}_0) = (\mathbf{B}'\mathbf{B})^{-1}\mathbf{B}'\mathbf{V}_n + o_p(1). \tag{9}$$

Proof: Since $\bar{\boldsymbol{\theta}}_n$ is consistent and is assumed to attain the minimum of Q_n over Ω, and since $\partial Q_n/\partial\theta_k$ exists near $\boldsymbol{\theta}_0$, it follows that for sufficiently large n

$$\frac{\partial Q_n}{\partial\theta_k}(\bar{\boldsymbol{\theta}}_n) = -\sum_{i=1}^{M} 2n\,\frac{N_i - np_i(\bar{\boldsymbol{\theta}}_n)}{N_i}\,\frac{\partial p_i}{\partial\theta_k}(\bar{\boldsymbol{\theta}}_n) = 0, \qquad k = 1, \ldots, m$$

with probability as near 1 as may be desired. Equivalently,

$$\sum_{i=1}^{M} \frac{N_i - np_i(\bar{\boldsymbol{\theta}}_n)}{N_i^{1/2}}\left(\frac{n}{N_i}\right)^{1/2}\frac{\partial p_i}{\partial\theta_k}(\bar{\boldsymbol{\theta}}_n) = 0, \qquad k = 1, \ldots, m. \quad (10)$$

We will apply the mean value theorem and the continuity theorem separately to the two factors in each summand of (10), remembering that $\bar{\boldsymbol{\theta}}_n \to \boldsymbol{\theta}_0(P)$ by Lemma 1. First,

$$\begin{aligned} N_i - np_i(\bar{\boldsymbol{\theta}}_n) &= N_i - np_i - n(p_i(\bar{\boldsymbol{\theta}}_n) - p_i) \\ &= N_i - np_i - n\sum_{k=1}^{m}\left[\frac{\partial p_i}{\partial\theta_k} + o_p(1)\right](\bar{\theta}_{nk} - \theta_{0k}). \end{aligned}$$

Combining this with

$$(np_i/N_i)^{1/2} = 1 + o_p(1) \quad (11)$$

establishes that $[N_i - np_i(\bar{\boldsymbol{\theta}}_n)]/N_i^{1/2}$ is the ith component of the M-vector

$$\mathbf{V}_n - \mathbf{B}n^{1/2}(\bar{\boldsymbol{\theta}}_n - \boldsymbol{\theta}_0) + o_p(1)n^{1/2}(\bar{\boldsymbol{\theta}}_n - \boldsymbol{\theta}_0) + o_p(1). \quad (12)$$

The second factor in (10) is similarly found to be

$$\left(\frac{n}{N_i}\right)^{1/2}\frac{\partial p_i}{\partial\theta_k}(\bar{\boldsymbol{\theta}}_n) = p_i^{-1/2}\frac{\partial p_i}{\partial\theta_k} + o_p(1).$$

Equation (10) therefore becomes, in vector form,

$$\mathbf{B}'\mathbf{V}_n - [\mathbf{B}'\mathbf{B} + o_p(1)]n^{1/2}(\bar{\boldsymbol{\theta}}_n - \boldsymbol{\theta}_0) = o_p(1). \quad (13)$$

Now $\mathbf{B}'\mathbf{B}$ is nonsingular by the rank m assumption, and since the determinant of a matrix is a continuous function of its elements,

$$\det(\mathbf{B}'\mathbf{B} + o_p(1)) \to \det(\mathbf{B}'\mathbf{B}) \neq 0(P).$$

Hence if A_n is the event that $\mathbf{B}'\mathbf{B} + o_p(1)$ is nonsingular, and χ_n

the indicator function of this event, then the probability of A_n under $G(\cdot|\boldsymbol{\theta}_0)$ approaches 1, $\chi_n \to 1(P)$, and $[\mathbf{B}'\mathbf{B} + o_p(1)]^{-1}\chi_n \to (\mathbf{B}'\mathbf{B})^{-1}(P)$. Applying $[\mathbf{B}'\mathbf{B} + o_p(1)]^{-1}\chi_n$ to both sides of (13) gives

$$[\mathbf{B}'\mathbf{B} + o_p(1)]^{-1}\chi_n\mathbf{B}'\mathbf{V}_n - n^{1/2}(\bar{\boldsymbol{\theta}}_n - \boldsymbol{\theta}_0)\chi_n = o_p(1)$$

which in turn implies the result of the theorem.

Theorem 3 yields immediate fruit, and (12) will produce a later harvest as well. Reviewing the proof, it is easy to see that (12) holds for *any* consistent estimator $\boldsymbol{\theta}_n$ of $\boldsymbol{\theta}$ in the form

$$\mathbf{V}_n(\boldsymbol{\theta}_n) = \mathbf{V}_n - \mathbf{B}n^{1/2}(\boldsymbol{\theta}_n - \boldsymbol{\theta}_0) + o_p(1)n^{1/2}(\boldsymbol{\theta}_n - \boldsymbol{\theta}_0) + o_p(1). \tag{14}$$

This is the central relation in the theory of chi-square tests, as it expresses $\mathbf{V}_n(\boldsymbol{\theta}_n)$ in terms of the standardized multinomial vector $\mathbf{V}_n$ and a separate term reflecting the effect of estimating $\boldsymbol{\theta}$. Notice that the third term on the right is $o_p(1)$ whenever $n^{1/2}(\boldsymbol{\theta}_n - \boldsymbol{\theta}_0)$ converges in distribution. We can now provide quick proofs of several important results.

The first of these is Fisher's solution to the question of the behavior of Pearson's statistic when $\boldsymbol{\theta}$ is estimated by $\bar{\boldsymbol{\theta}}_n$. Substituting (9) into (14), we see that

$$\mathbf{V}_n(\bar{\boldsymbol{\theta}}_n) = (\mathbf{I}_M - \mathbf{B}(\mathbf{B}'\mathbf{B})^{-1}\mathbf{B}')\mathbf{V}_n + o_p(1).$$

Asymptotic normality for $\mathbf{V}_n$ was immediate (see (7)). By the continuity theorem and the result from Section 2 on linear transformations of multivariate normal variables, it follows that under $G(\cdot|\boldsymbol{\theta}_0)$,

$$\mathbf{V}_n(\bar{\boldsymbol{\theta}}_n) \xrightarrow{\mathscr{D}} N_M(\mathbf{0}, \boldsymbol{\Sigma})$$
$$\begin{aligned}\boldsymbol{\Sigma} &= (\mathbf{I}_M - \mathbf{B}(\mathbf{B}'\mathbf{B})^{-1}\mathbf{B}')(\mathbf{I}_M - \mathbf{q}\mathbf{q}')(\mathbf{I}_M - \mathbf{B}(\mathbf{B}'\mathbf{B})^{-1}\mathbf{B}') \\ &= \mathbf{I}_M - \mathbf{q}\mathbf{q}' - \mathbf{B}(\mathbf{B}'\mathbf{B})^{-1}\mathbf{B}'.\end{aligned}$$

The last equality is a consequence of the important relation $\mathbf{q}'\mathbf{B} = 0$, which holds because $\sum_{i=1}^M p_i = 1$ implies that $\sum_{i=1}^M \partial p_i/\partial\theta_k = 0$ for each k. The limiting null distribution of any statistic $\mathbf{V}_n(\bar{\boldsymbol{\theta}}_n)'\mathbf{C}\mathbf{V}_n(\bar{\boldsymbol{\theta}}_n)$ is now given by Theorem 1. In particular, the Pearson statistic $P_n(\bar{\boldsymbol{\theta}}_n) = \mathbf{V}_n(\bar{\boldsymbol{\theta}}_n)'\mathbf{V}_n(\bar{\boldsymbol{\theta}}_n)$ has the distribution of $\sum_{i=1}^M \lambda_i Z_i^2$ where λ_i are the characteristic roots of $\boldsymbol{\Sigma}$. A bit of matrix multiplication will show that $\mathbf{q}\mathbf{q}'$ and $\mathbf{B}(\mathbf{B}'\mathbf{B})^{-1}\mathbf{B}'$ are symmetric idempotent matrices, that is, orthogonal projections. Moreover, we just saw that they are orthogonal to each other. Because $\mathbf{qq}'$ has rank 1 and $\mathbf{B}(\mathbf{B}'\mathbf{B})^{-1}\mathbf{B}'$ has rank m by assumption, $\mathbf{\Sigma}$ is an orthogonal projection of rank $M - m - 1$. So its characteristic roots are $M - m - 1$ 1's and $m + 1$ 0's, and the limiting null distribution of $P_n(\bar{\boldsymbol{\theta}}_n)$ is $\chi^2(M - m - 1)$. Notice especially that this is true for any $\boldsymbol{\theta}_0$ in Ω, even though $\mathbf{\Sigma}$ varies with $\boldsymbol{\theta}_0$. This is the famous "subtract one degree of freedom for each parameter estimated" result.

Now $\bar{\boldsymbol{\theta}}_n$ is often not the most convenient available estimator of $\boldsymbol{\theta}$. In testing fit to the univariate normal family with $\boldsymbol{\theta} = (\mu, \sigma)$, for example, the cell probability for a cell $E_i = (a_{i-1}, a_i]$ is

$$p_i(\mu, \sigma) = \Phi\left(\frac{a_i - \mu}{\sigma}\right) - \Phi\left(\frac{a_{i-1} - \mu}{\sigma}\right),$$

where Φ is the standard normal d.f. The equations (10) have no closed-form solution, nor do the yet more complicated equations $\partial P_n(\boldsymbol{\theta})/\partial\theta_k = 0$ defining the minimum chi-square estimator. A visit to your local computing center will uncover library programs for evaluating Φ and iteratively solving the equations (10). Nonetheless, it is hard to ignore the universally used sample mean and variance, $\hat{\boldsymbol{\theta}}_n = (\bar{X}, s)$. What will befall us if we use these estimators in the Pearson statistic instead of $\bar{\boldsymbol{\theta}}_n$? To answer this question, we must discover the large-sample behavior of $\hat{\boldsymbol{\theta}}_n$ and then consult (14).

This is best done in greater generality. The sample mean and standard deviation (taking $s^2 = \sum_{j=1}^n (X_j - \bar{X})^2/n$) form the *maximum likelihood estimator* (MLE) of $\boldsymbol{\theta} = (\mu, \sigma)$ in the univariate normal family. In general, the MLE $\hat{\boldsymbol{\theta}}_n = \hat{\boldsymbol{\theta}}_n(\mathbf{X}_1, \ldots, \mathbf{X}_n)$ of $\boldsymbol{\theta}$ is defined as any value of $\boldsymbol{\theta}$ maximizing the joint density function of the observations considered as a function of $\boldsymbol{\theta}$ for given $\mathbf{X}_1, \ldots, \mathbf{X}_n$. This recipe for a general method of estimating parameters is another of Fisher's contributions. It is intuitively forceful, estimating $\boldsymbol{\theta}$ to be the value making the actually observed $\mathbf{X}_1, \ldots, \mathbf{X}_n$ "most probable." More satisfying to the perverse theoretician, the MLE is guaranteed to have good properties in large samples. Specifically, suppose that the $\mathbf{X}_j$ are independent with common density function $g(\cdot|\boldsymbol{\theta}_0)$. Then under reasonable smoothness conditions,

$$n^{1/2}(\hat{\boldsymbol{\theta}}_n - \boldsymbol{\theta}_0) = \mathbf{J}(\boldsymbol{\theta}_0)^{-1}n^{-1/2}\sum_{j=1}^{n}\frac{\partial \log g(\mathbf{X}_j|\boldsymbol{\theta}_0)}{\partial\boldsymbol{\theta}} + o_p(1). \qquad (15)$$

Here $\partial \log g/\partial \boldsymbol{\theta}$ is the m-vector of partial derivatives with respect to $\theta_1, \ldots, \theta_m$ and $\mathbf{J}(\boldsymbol{\theta})$ is the $m \times m$ matrix with (k, l)th component

$$E_{\boldsymbol{\theta}}\left[\frac{\partial \log g(\mathbf{X}|\boldsymbol{\theta})}{\partial \theta_k} \frac{\partial \log g(\mathbf{X}|\boldsymbol{\theta})}{\partial \theta_l}\right].$$

It follows from (15) by the multivariate central limit theorem that

$$n^{1/2}(\hat{\boldsymbol{\theta}}_n - \boldsymbol{\theta}_0) \xrightarrow{\mathscr{D}} N_m(0, \mathbf{J}(\boldsymbol{\theta}_0)^{-1}). \tag{16}$$

The matrix $\mathbf{J}(\boldsymbol{\theta})$ is called the *information matrix* for the family $g(\mathbf{x}|\boldsymbol{\theta})$. The inverse $\mathbf{J}(\boldsymbol{\theta}_0)^{-1}$ is the "smallest possible" covariance matrix for the limiting distribution of an estimator of $\boldsymbol{\theta}$, in several specific senses which this is not the place to specify. Thus (16) says roughly that the MLE has the tightest possible concentration about the true $\boldsymbol{\theta}_0$ in large samples. This is called *asymptotic efficiency* of the MLE.

In the light of this pleasing result, it would be very intelligent, if we wish to estimate $\boldsymbol{\theta}$ from cell frequencies, to apply the MLE recipe to the indicator variables $\boldsymbol{\delta}_1, \ldots, \boldsymbol{\delta}_n$ indicating into which cells $\mathbf{X}_1, \ldots, \mathbf{X}_n$ fall. A bit of work shows that the information matrix in this case is $\mathbf{B}'\mathbf{B}$, and that (15) reduces to (9). *The minimum chi-square and minimum modified chi-square and maximum likelihood estimators are all asymptotically equivalent ways of estimating* $\boldsymbol{\theta}$ *from the cell frequencies.* That's aesthetically satisfying.

Having summed up half a century of hard work on MLE's in one paragraph, we can now substitute (15) into (14). Here $\hat{\boldsymbol{\theta}}_n$ is the MLE of $\boldsymbol{\theta}$ from the ungrouped observations $\mathbf{X}_1, \ldots, \mathbf{X}_n$, not the less efficient MLE based on the cell frequencies. Fortune is with us. The first term in (14), namely $\mathbf{V}_n$, was expressed at the beginning of this section as a sum of n terms, one for each $\mathbf{X}_j$. The second term, namely (15), has the same form. And the rest of (14) is $o_p(1)$. So we obtain from the first two terms a sum which is asymptotically normal by the multivariate central limit theorem. A computation of covariances gives specifically that

$$\mathbf{V}_n(\hat{\boldsymbol{\theta}}_n) \xrightarrow{\mathscr{D}} N_M(\mathbf{0}, \boldsymbol{\Sigma})$$
$$\boldsymbol{\Sigma} = \mathbf{I}_M - \mathbf{q}\mathbf{q}' - \mathbf{B}\mathbf{J}^{-1}\mathbf{B}'.$$

Therefore the limiting null distribution of $P_n(\hat{\boldsymbol{\theta}}_n)$ is that of $\sum_{i=1}^{M} \lambda_i Z_i^2$, where λ_i are the characteristic roots of $\boldsymbol{\Sigma}$.

Now $\mathbf{BJ}^{-1}\mathbf{B}'$ has the same rank m as does $\mathbf{B}$, and therefore has as its range the range of $\mathbf{B}$. Since $\mathbf{qq}'$ is an orthogonal projection of rank 1 orthogonal to $\mathbf{B}$, the characteristic roots of $\boldsymbol{\Sigma}$ include $M - m - 1$ 1's ($\boldsymbol{\Sigma}$ acts as the identity in directions orthogonal to the direct sum of the ranges of $\mathbf{B}$ and $\mathbf{qq}'$) and one 0 ($\boldsymbol{\Sigma}$ acts as zero on the range of $\mathbf{qq}'$). The remaining roots $\lambda_1, \ldots, \lambda_m$ reflect the fact that $\boldsymbol{\Sigma}$ acts as $\mathbf{I}_M - \mathbf{BJ}^{-1}\mathbf{B}'$ on the range of $\mathbf{B}$. One version of the "efficiency" of the MLE is that $\hat{\boldsymbol{\theta}}_n$ is asymptotically preferable to $\bar{\boldsymbol{\theta}}_n$ in the sense that $\mathbf{J} - \mathbf{B}'\mathbf{B}$ is nonnegative definite. From this it can be shown by matrix mangling that $0 \leqslant \lambda_i < 1$, and $0 < \lambda_i < 1$ except in the unusual case when $\mathbf{J} - \mathbf{B}'\mathbf{B}$ fails to be positive definite. The λ_i of course depend on $\boldsymbol{\theta}_0$, as well as on the specific hypothesized family $g(\cdot|\boldsymbol{\theta})$.

We have now reached the second major consequence of (14). The statistic $P_n(\hat{\boldsymbol{\theta}}_n)$ has as its limiting null distribution the distribution of

$$\chi^2(M - m - 1) + \sum_{i=1}^{m} \lambda_i Z_i^2. \tag{17}$$

This is *not* a chi-square distribution. What is worse, the distribution varies with $\boldsymbol{\theta}_0$, so there is no single limiting distribution across the composite null hypothesis. Since $0 \leqslant \lambda_i < 1$ for all $\boldsymbol{\theta}_0$, it is at least true that critical points of (17) lie between those of $\chi^2(M - m - 1)$ and $\chi^2(M - 1)$. When there are many cells and few parameters, these bounds are close together. But we cannot without care follow such natural paths as the use of $\bar{X}$ and s in the Pearson statistic to test for normality.

After Chernoff and Lehmann [2] obtained the result (17) in 1954, statistical theory produced a variety of ways of escape. One is suggested immediately by Theorem 2: Compute a generalized inverse of $\boldsymbol{\Sigma}$ and use the corresponding quadratic form. It is easy to see from our previous study of $\boldsymbol{\Sigma}$ that $\boldsymbol{\Sigma}$ has rank $M - 1$ and

$$\boldsymbol{\Sigma}^{-} = (\mathbf{I}_M - \mathbf{BJ}^{-1}\mathbf{B}')^{-1}$$

whenever $\mathbf{J} - \mathbf{B}'\mathbf{B}$ is positive definite. If now

$$\mathbf{C}_n = (\mathbf{I}_M - \mathbf{B}(\hat{\boldsymbol{\theta}}_n)\mathbf{J}(\hat{\boldsymbol{\theta}}_n)^{-1}\mathbf{B}(\hat{\boldsymbol{\theta}}_n)')^{-1}$$

then $\mathbf{C}_n \to \boldsymbol{\Sigma}^-(P)$ and

$$\mathbf{V}_n(\hat{\boldsymbol{\theta}}_n)'\mathbf{C}_n\mathbf{V}_n(\hat{\boldsymbol{\theta}}_n) \to \chi^2(M - 1)$$

under $G(\cdot|\boldsymbol{\theta})$ for any $\boldsymbol{\theta}$ in Ω. This statistic is not as hard to compute as may appear, as will be shown by example in Section 7. This statistic was first studied by Rao and Robson [5], but without the supporting theory. Rao and Robson present $\mathbf{C}_n$ in the form

$$\mathbf{C}_n = \mathbf{I}_M + \mathbf{B}(\hat{\boldsymbol{\theta}}_n)[\mathbf{J}(\hat{\boldsymbol{\theta}}_n) - \mathbf{B}(\hat{\boldsymbol{\theta}}_n)'\mathbf{B}(\hat{\boldsymbol{\theta}}_n)]^{-1}\mathbf{B}(\hat{\boldsymbol{\theta}}_n)'$$

which makes it clear that the new statistic $\mathbf{V}_n(\hat{\boldsymbol{\theta}}_n)'\mathbf{C}_n\mathbf{V}_n(\hat{\boldsymbol{\theta}}_n)$ is the Pearson statistic plus a second quadratic form. Challenge: prove that the two expressions given for $\mathbf{C}_n$ are equivalent.

If $P_n(\hat{\boldsymbol{\theta}}_n)$ can be built up to reach $\chi^2(M - 1)$, it can also be chopped down to $\chi^2(M - m - 1)$. Since $\mathbf{B}(\mathbf{B}'\mathbf{B})^{-1}\mathbf{B}'$ is the orthogonal projection onto the range of $\mathbf{B}$, you should be able to show that $\mathbf{V}_n(\hat{\boldsymbol{\theta}}_n)'\mathbf{D}(\hat{\boldsymbol{\theta}}_n)\mathbf{V}_n(\hat{\boldsymbol{\theta}}_n)$ has the $\chi^2(M - m - 1)$ limiting null distribution, where

$$\mathbf{D}(\boldsymbol{\theta}) = \mathbf{I}_M - \mathbf{B}(\boldsymbol{\theta})[\mathbf{B}'(\boldsymbol{\theta})\mathbf{B}(\boldsymbol{\theta})]^{-1}\mathbf{B}'(\boldsymbol{\theta}).$$

This result does not even depend on the use of $\hat{\boldsymbol{\theta}}_n$; $\bar{\boldsymbol{\theta}}_n$ and most other estimators of $\boldsymbol{\theta}$ give the same result. But the price of such generality is inefficiency. Simulations suggest that the $\mathbf{D}(\hat{\boldsymbol{\theta}}_n)$ statistic often has low power (that is, little ability to detect that the null hypothesis is false). The $\mathbf{C}_n$ statistic, on the other hand, is usually more powerful than the Pearson statistic. It deserves consideration as a standard chi-square test for goodness of fit.

5. CONTINGENCY TABLES

The use of chi-square statistics for testing fit is based on creating a set of multinomial observations by counting cell frequencies. Because only cell frequencies are used in the tests, some information is lost. There are other classes of tests of fit which are generally more powerful than chi-square tests, though none so flexible and widely applicable. There are, however, situations in which multinomial observations arise naturally. In such cases, chi-square tests are the natural large sample tests. A common instance is a *contingency table*: sample units are categorized according to two or more variables with the intent of discovering the relationship between the variables. The data consist of the frequencies of sample units in all possible cross-classifications. Here is the layout of a $2 \times s$ contingency table, with the notation used for the cell frequencies.

$$
\begin{array}{|c|c|c|c|l}
\hline
N_{11} & N_{12} & \cdots & N_{1s} & N_{1\cdot} \\
\hline
N_{21} & N_{22} & \cdots & N_{2s} & N_{2\cdot} \\
\hline
\multicolumn{1}{c}{N_{\cdot 1}} & \multicolumn{1}{c}{N_{\cdot 2}} & \multicolumn{1}{c}{} & \multicolumn{1}{c}{N_{\cdot s}} &
\end{array}
\qquad (18)
$$

We have used the common notation in which a dot replaces an index when the frequencies are summed over the full range of that index. Thus $N_{\cdot j}$ is the jth column sum, the total number of units which fell in category j for the column variable. For simplicity, this $2 \times s$ table will be the focus of our discussion, though the conclusions are generally valid.

What is the proper probability model for these data? *The model must reflect the way in which the data were collected.* There are several different sampling procedures which could lead to the table (18). A single random sample of size n might be selected, then categorized in two ways. For example, a random sample of persons being treated for cancer might be classified by sex (2 categories) and type of cancer (s categories). Call this Model A. Under Model A the cell frequencies N_{ij} have a single $2s$-nomial distribution. The marginal frequencies are all random, and satisfy

$$
N_{1\cdot} + N_{2\cdot} = \sum_{j=1}^{s} N_{\cdot j} = \sum_{i=1}^{2} \sum_{j=1}^{s} N_{ij} = n. \qquad (19)
$$

Table (18) might also result from selecting two independent random samples, of male cancer patients and female cancer patients separately, then categorizing each patient by type of cancer. Under this Model B, table (18) contains two independent s-nomial distributions. Although (19) still holds, $N_{1\cdot}$ and $N_{2\cdot}$ are no longer random, for they are the sample sizes chosen by the experimenter. Model C reverses the roles of the variables: choose independent random samples of patients under treatment for each of s types of cancer, then categorize each by sex. Here there are s independent binomials, and the $N_{\cdot j}$ are nonrandom sample sizes.

All three models for table (18) are sets of independent multinomial observations. Chi-square methods provide tests of hypotheses concerning the cell probabilities in any such setting. This is a generalization of the situation arising in tests of fit, where only a single multinomial sample was available, but the theory of chi-square tests follows much the same line.

Hypotheses for these models are stated in terms of the cell probabilities p_{ij} for the sampled population. Each model imposes different constraints on the p_{ij}. Model A requires only that

$$\sum_{i=1}^{2} \sum_{j=1}^{s} p_{ij} = 1 \tag{20}$$

(and of course that $0 \leqslant p_{ij} \leqslant 1$ for all i and j). Model B states that

$$\begin{aligned} p_{1\cdot} &= \sum_{j=1}^{s} p_{1j} = 1 \\ p_{2\cdot} &= \sum_{j=1}^{s} p_{2j} = 1, \end{aligned} \tag{21}$$

since in this case two independent s-nomials are observed. Model C assumes instead that $p_{\cdot j} = p_{1j} + p_{2j} = 1$ for each j. The most common hypotheses (and the only ones we will consider) formalize the statement that there is no connection between the two categorizations—in the example, no connection between the sex of a cancer patient and the type of cancer under treatment. In Model A, this is the hypothesis of *independence*,

$$H_A : p_{ij} = p_{i\cdot}p_{\cdot j} \qquad i = 1, 2 \text{ and } j = 1, \ldots, s. \tag{22}$$

In Model B, the hypothesis is that of *two identical s-nomial distributions*,

$$H_B : p_{1j} = p_{2j} \qquad j = 1, \ldots, s. \tag{23}$$

For Model C, no connection between categorizations is expressed as the hypothesis of *s identical binomial distributions*,

$$H_C : p_{11} = p_{12} = \cdots = p_{1s}.$$

In all of this, our concern has been simply to translate the sampling design and the question to be asked of the data into a mathematical model. This process is often less clear and more controversial than the theory which follows from the model selected. There are, for example, yet other models for the data of table (18). These models assume that the data arise not as random samples from a large population, but from experimental randomization of a finite set of experimental units. In the randomization analog of Model B, n units (say lab rats) are available, of which $N_{1\cdot}$ are assigned to Treatment 1 and $N_{2\cdot}$ to Treatment 2 by random allocation. The response of each rat falls into one of s categories, and N_{ij} is the number of rats receiving Treatment i with response j resulting. Just as in Model B, (19) holds, $N_{1\cdot}$ and $N_{2\cdot}$ are fixed, and the hypothesis to be tested is that of equal response distributions. But under the null hypothesis of "no treatment effect," the number $N_{\cdot j}$ of rats showing response j is a nonrandom characteristic of the particular set of rats used in the experiment. The N_{ij} are dependent, and each has a hypergeometric distribution. Thus multinomial models do not apply.

Now it turns out that under Model B the *conditional* null distribution of the N_{ij} given the observed values of $N_{\cdot j}$ ($j = 1, \ldots, s$) is exactly the null distribution of the N_{ij} under the randomization model just described. Because such experimental randomization is common practice in many fields of work, it has been the historical pattern to argue that the randomization models are "exact" and the multinomial models (and therefore the chi-square tests) are valid only when interpreted conditionally. But wait—since the randomization model considers only the fixed set of units actually involved in the experiment, any inference based on that model can apply only to those particular units. If our n rats are in some way atypical of rats-in-general, this will influence the outcome of the experiment, and no conclusions can be drawn for rats-in-general. In practice, we argue or assume that our particular units are representative of some larger population. That is, we commonly act as if we had samples from a population of interest. What is more, steps are often taken to justify this assumption—we cannot sample the population of all rats or all cancer patients, but we can select our experimental units at random from a large pool of accessible units. In such a case neither the randomization nor the multinomial models are strictly appropriate, but the multinomial models do represent the conditions which the selection and allocation of units aim to attain.

This discussion is not at all a digression. It is rather a paradigm of the features which distinguish areas of applied mathematics (such as statistics) from mathematics-for-its-own-sake. It is time, however, to assume that one of the multinomial models adequately describes the data and to turn to tests of H_A, H_B or H_C. If we wish to detect any deviation from the null hypothesis (not just deviations in some specified direction), an omnibus test is in order. And if the sample sizes are moderately large, chi-square methods provide such tests.

We will follow the pattern of Section 4, denoting the unknown vector of parameters by $\boldsymbol{\theta}$. In the contingency table case, $\boldsymbol{\theta}$ consists of a set of cell probabilities which determine all of the p_{ij} when combined with the constraints imposed by the model and by the null hypothesis. Under Model B, for example, we will take $\boldsymbol{\theta} = (p_{11}, \ldots, p_{1,s-1})'$, since (21) and (23) then determine the complete set of cell probabilities. The estimators, test statistics, and limiting distributions discussed below do not depend on this particular choice of $s - 1$ cell probabilities for $\boldsymbol{\theta}$, but the dimension $m = s - 1$ of $\boldsymbol{\theta}$ is a consequence of the model and the hypothesis.

The probability function in Model B is

$$\frac{N_1!}{N_{11}! \cdots N_{1s}!} \prod_{j=1}^{s} p_{1j}^{N_{1j}} \frac{N_2!}{N_{21}! \cdots N_{2s}!} \prod_{j=1}^{s} p_{2j}^{N_{2j}}.$$

To estimate $\boldsymbol{\theta}$ by the maximum likelihood method, express each p_{ij}

as $p_{ij}(\boldsymbol{\theta})$ and the probability function for given N_{ij} as a function $L(\boldsymbol{\theta})$ of $\boldsymbol{\theta}$. Then solving

$$\frac{\partial \log L(\boldsymbol{\theta})}{\partial \theta_k} = \frac{N_{1k}}{p_{1k}} - \frac{N_{1s}}{p_{1s}} + \frac{N_{2k}}{p_{1k}} - \frac{N_{2s}}{p_{1s}} = 0 \qquad k = 1, \ldots, s-1$$

(recall $\theta_k = p_{1k}$) produces the MLE

$$\hat{p}_{1k} = \frac{N_{1k} + N_{2k}}{n} = \frac{N_{\cdot k}}{n} \qquad k = 1, \ldots, s-1.$$

This is the "obvious" estimator of p_{1k} under H_B, namely the overall relative frequency of the kth response in the two samples. Another way to describe $\hat{p}_{1k}$ is as the weighted arithmetic mean of the relative frequencies $N_{1k}/N_{1\cdot}$ and $N_{2k}/N_{2\cdot}$ of the kth response in the separate samples. The Pearson chi-square statistic for two independent s-nomials is

$$\sum_{j=1}^{s} \frac{[N_{1j} - np_{1j}(\hat{\boldsymbol{\theta}})]^2}{np_{1j}(\hat{\boldsymbol{\theta}})} + \sum_{j=1}^{s} \frac{[N_{2j} - np_{2j}(\hat{\boldsymbol{\theta}})]^2}{np_{2j}(\hat{\boldsymbol{\theta}})}$$

$$= \sum_{i=1}^{2} \sum_{j=1}^{s} \frac{[N_{ij} - N_{i\cdot}N_{\cdot j}/n]^2}{N_{i\cdot}N_{\cdot j}/n}.$$

When the p_{ij} are known, the corresponding statistic would have $(s-1) + (s-1) = 2s - 2$ degrees of freedom. Here, however, $m = s - 1$ parameters were estimated by the multinomial MLE method, so since $(2s-2) - (s-1) = s - 1$, the limiting null distribution of this statistic is $\chi^2(s-1)$.

If the data of table (18) arose from a single random sample (Model A), the probability function is

$$\prod_{i=1}^{2} \prod_{j=1}^{s} \frac{n!}{N_{ij}!} p_{ij}^{N_{ij}}.$$

The unknown parameter can be taken to be $\boldsymbol{\theta} = (p_{11}, \ldots, p_{1s})'$ since (20) and (22) then determine the full set of cell probabilities. Once again other choices of $\boldsymbol{\theta}$ are possible but all have $m = s$. Computing the MLE of $\boldsymbol{\theta}$ gives the natural estimator for Model A under H_A, namely

$$\hat{p}_{1k} = \frac{N_{1\cdot}}{n} \frac{N_{\cdot k}}{n}.$$

(Compare (22) to see why this is the natural estimator.) The Pearson chi-square statistic for this single $2s$-nomial model is

$$\sum_{i=1}^{2}\sum_{j=1}^{s}\frac{[N_{ij}-np_{ij}(\hat{\boldsymbol\theta})]^2}{np_{ij}(\hat{\boldsymbol\theta})}=\sum_{i=1}^{2}\sum_{j=1}^{s}\frac{[N_{ij}-N_{i\cdot}N_{\cdot j}/n]^2}{N_{i\cdot}N_{\cdot j}/n}$$

and has $2s - m - 1 = s - 1$ degrees of freedom.

Look closely. *The Pearson chi-square statistics for testing H_A in* Model A *and for testing H_B in* Model B *are identical, and have the same $\chi^2(s-1)$ limiting null distribution.* And of course the same statistic results from testing H_C in Model C. This serendipitous outcome depends very much on the fact that maximum likelihood estimation was used. As Section 4 proclaimed, asymptotically equivalent statistics can be obtained by using either the minimum chi-square or the minimum modified chi-square method to estimate $\boldsymbol\theta$. But only asymptotically equivalent. Let us apply the minimum modified chi-square method to Model B. We must choose $\boldsymbol\theta = (p_{11}, \ldots, p_{1,s-1})'$ to minimize the modified chi-square statistic

$$\sum_{j=1}^{s}\frac{[N_{1j}-N_{1\cdot}p_{1j}(\boldsymbol\theta)]^2}{N_{1j}}+\sum_{j=1}^{s}\frac{[N_{2j}-N_{2\cdot}p_{2j}(\boldsymbol\theta)]^2}{N_{2j}}.$$

Differentiation followed by a short, ugly calculation shows that the minimum modified chi-square estimator of p_{1k} is proportional to the weighted *harmonic* mean of the relative frequencies $N_{1k}/N_{1\cdot}$ and $N_{2k}/N_{2\cdot}$ for the kth response. While not entirely outrageous, this is surely less appealing than the MLE result. In Model A, the situation is worse: the equations arising from differentiating the modified chi-square statistic are nonlinear, and have no closed-form solution. The Pearson statistics for Models A and B when minimum modified chi-square estimators are used are *not* identical. The minimum chi-square estimators have no explicit expressions in either model, and again the chi-square statistics differ. No wonder the MLE is always used for contingency tables.

There is a pattern to the use of these latter estimation methods in hypothesis testing for independent multinomial observations. The minimum modified chi-square method produces a set of *linear* equations to be solved for the estimated parameters whenever the hypothesis is linear in the cell probabilities. This was true of H_B but not of H_A. In many situations it is easier to compute minimum modified chi-square estimators than the MLE's. Minimum chi-square estimators, on the other hand, can rarely be obtained in closed form and are seldom used.

6. A FURTHER RANGE

This survey of chi-square tests has entirely ignored several areas of considerable interest to users of these tests. Computers make it feasible to obtain the exact distributions of the test statistics in small samples, both for use and for assessment of the accuracy of the chi-square approximations. The relative power of the tests can be studied either by calculation and simulation or, in large samples, by various mathematical devices. I have chosen to restrict this essay to the study of large sample distribution theory under the null hypothesis. Even here there is a further range of theory which both opens up new possibilities for the user and illustrates the use of increasingly sophisticated mathematics in statistical theory.

We have been assuming almost without reflection that the number of cells M in a chi-square test of fit remains fixed as the sample size increases, and that the cells E_i are fixed without regard to the data. Neither assumption is necessarily realistic as a description of statistical practice. It is common to use more cells when blessed with a larger sample, and equally common (though less publicly admitted) to move the cells to the data. What are the consequences of incorporating these innovations in the chi-square statistics of Section 4?

Increasing the number of cells M as the sample size n increases has radical consequences. When M grows with n, the Pearson statistic for testing fit to a completely specified distribution has a *normal*, not a chi-square limiting null distribution when properly standardized. This is in accord with intuition, since the $\chi^2(M-1)$ limiting distribution of the Pearson statistic approaches normality as $M \to \infty$. (Apply the central limit theorem to $\sum_1^{M-1} Z_i^2$.) One expects that the limiting null distribution when parameters are estimated will also be normal, with a different standardization perhaps required. No proof of this has been given.

The second innovation, use of data-dependent cells, has been better studied and is finding its way increasingly into practical use. Suppose then that the cells E_i are replaced by

$$E_{in}(\mathbf{X}_1, \ldots, \mathbf{X}_n) \qquad i = 1, \ldots, M$$

in the general chi-square statistic (6) of Section 4. For simplicity we consider only univariate observations X_j and cells $E_{in} = (a_{i-1,n}, a_{in}]$ which are intervals with endpoints $a_{in} = a_{in}(X_1, \ldots, X_n)$. It is only reasonable to demand that the random cells settle down as the sample size increases,

$$a_{in} \to a_{i0} = a_{i0}(\boldsymbol{\theta}_0)(P) \qquad \text{under} \qquad G(\cdot|\boldsymbol{\theta}_0).$$

An example of useful data-dependent cell boundaries is $a_{in} = \bar{X}_n + c_i s_n$ in testing fit to the univariate normal family. The sample mean $\bar{X}_n$ moves the cells to the data, and the sample standard deviation s_n adjusts the cell widths to the dispersion of the data. Here $a_{i0}(\mu, \sigma) = \mu + c_i\sigma$ are the limiting cell boundaries.

If $\mathbf{a}_n$ denotes the vector of cell boundaries ($a_{0n} \equiv -\infty, a_{Mn} \equiv +\infty$), then the "cell probabilities" under the null hypothesis are now

$$p_i(\boldsymbol{\theta}, \mathbf{a}_n) = \int_{a_{i-1,n}}^{a_{in}} dG(x|\boldsymbol{\theta}) = G(a_{in}|\boldsymbol{\theta}) - G(a_{i-1,n}|\boldsymbol{\theta}).$$

The M-vector of standardized cell frequencies becomes $\mathbf{V}_n(\boldsymbol{\theta}, \mathbf{a}_n)$ with ith component

$$\frac{N_i(\mathbf{a}_n) - np_i(\boldsymbol{\theta}, \mathbf{a}_n)}{[np_i(\boldsymbol{\theta}, \mathbf{a}_n)]^{1/2}}$$

where $N_i(\mathbf{a}_n)$ is the number of $X_1, \ldots, X_n$ in E_{in}. But *the cell frequencies $N_i(\mathbf{a}_n)$ are no longer multinomial*, since the cell boundaries are dependent on the observations X_j being counted. The central mathematical hurdle of establishing asymptotic normality of $\mathbf{V}_n(\boldsymbol{\theta}_n, \mathbf{a}_n)$ for estimators $\boldsymbol{\theta}_n$ and random cell boundaries $\mathbf{a}_n$ is now much more difficult. Fortunately, there is an elegant modern technique leading to a pleasing result.

The pleasing result is that the asymptotic distribution of $\mathbf{V}_n(\boldsymbol{\theta}_n, \mathbf{a}_n)$ under $G(\cdot\,|\boldsymbol{\theta}_0)$ is the same as that of $\mathbf{V}_n(\boldsymbol{\theta}_n, \mathbf{a}_0)$. That is, *the asymptotic behavior of any random-cell chi-square statistic is exactly that of the same statistic using the limiting cell boundaries* a_{i0}. Speaking roughly, the use of data-dependent cells has no effect. The naive user who moves his cells to the data is safe. Actually, the dependence of the limiting cells on the (unknown) true $\boldsymbol{\theta}_0$ complicates these rough conclusions slightly. But any of the statistics in Section 4 which has a $\boldsymbol{\theta}_0$-free limiting null distribution using fixed cells has that same distribution using any set of converging random cells.

What of the promised elegant modern technique? Since $\mathbf{V}_n(\boldsymbol{\theta}_n, \mathbf{a}_0)$ is just our old acquaintance $\mathbf{V}_n(\boldsymbol{\theta}_n)$ for a particular set of fixed cells, let us use the latter notation. We must show that under $G(\cdot\,|\boldsymbol{\theta}_0)$

$$\mathbf{V}_n(\boldsymbol{\theta}_n, \mathbf{a}_n) - \mathbf{V}_n(\boldsymbol{\theta}_n) = o_p(1).$$

Since both $p_i(\boldsymbol{\theta}_n, \mathbf{a}_n)$ and $p_i(\boldsymbol{\theta}_n)$ converge in probability to $p_i(\boldsymbol{\theta}_0)$ whenever $\boldsymbol{\theta}_n$ is a consistent sequence of estimators and p_i is continuous, their presence in the denominators can be ignored. We need only prove that the expression

$$n^{-1/2}[N_i(\mathbf{a}_n) - np_i(\boldsymbol{\theta}_n, \mathbf{a}_n)] - n^{-1/2}[N_i - np_i(\boldsymbol{\theta}_n)] \tag{24}$$

is $o_p(1)$.

Introduce the *empirical distribution function*

$$G_n(t) = n^{-1}\{\text{Number of } X_1, \ldots, X_n\colon X_j \leqslant t\}.$$

This is the natural estimator of the common d.f. of the X_j. It increases from 0 to 1 in jumps of $1/n$ at each observation. At any fixed t, $G_n(t)$ is a multiple of a binomial random variable with success probability $G(t\,|\boldsymbol{\theta}_0)$ when this is the true d.f. of the X_j. So the *empirical distribution function process*

$$W_n(t) = n^{1/2}\{G_n(t) - G(t\,|\boldsymbol{\theta}_0)\}$$

has a normal limiting distribution for each fixed t. Since

$$n^{-1/2}N_i(\mathbf{a}_n) = n^{1/2}\{G_n(a_{in}) - G_n(a_{i-1,n})\}$$
$$p_i(\boldsymbol{\theta}, \mathbf{a}_n) = G(a_{in}|\boldsymbol{\theta}) - G(a_{i-1,n}|\boldsymbol{\theta}),$$

there is some hope of expressing (24) in terms of W_n and using the convergence properties of that process to achieve our goal.

In Section 3 we saw that convergence in distribution for random variables amounted to describing a random variable $\mathbf{Y}_n$ by a probability measure F_n on R^p and defining convergence as convergence of the measures $F_n(A)$ of all Borel sets A having boundaries with measure 0 under the limiting distribution. This development extends at once to more general spaces than R^p. Indeed, such an extension is the primary topic of Billingsley's book [1] which was cited in Section 3. Now a stochastic process such as W_n is a *random function*—a function of the real variable t which varies with the underlying probability mechanism generating the X_j. Just as a random variable can be identified with a probability measure on R^p, so a random function can be identified with a probability measure on a suitable function space. Convergence in distribution for processes then has the same definition as convergence in distribution for random variables. This viewpoint, adopted from functional analysis, has become a standard tool of statistical large sample theory. Billingsley's book is a basic exposition, restricted to metric spaces, Borel σ-fields, and random functions of a single real variable.

It turns out that $W_n(t)$ does converge in distribution to a process $W_0(t)$ which is a variation of Brownian motion, one of the most familiar stochastic processes. This is an analog of the central limit theorem. Of this powerful result we need only two details: $W_0(t)$ is continuous with probability 1, and the function space on which W_n and W_0 are probability measures has the property that convergence to a continuous limit function is always uniform.

The machinery to crush (24) is now assembled. Arithmetic shows that (24) is

$$\begin{aligned}&\{W_n(a_{in}) - W_n(a_{i0})\} - \{W_n(a_{i-1,n}) - W_n(a_{i-1,0})\}\\ &\quad + n^{1/2}\{p_i(\mathbf{a}_n, \boldsymbol{\theta}_0) - p_i(\mathbf{a}_n, \boldsymbol{\theta}_n)\} - n^{1/2}\{p_i(\mathbf{a}_0, \boldsymbol{\theta}_0) - p_i(\mathbf{a}_0, \boldsymbol{\theta}_n)\}. \quad (25)\end{aligned}$$

Applying the mean value theorem to the last two terms in (25) gives

$$\left[\frac{\partial p_i}{\partial \theta}(\boldsymbol{\theta}_n^*, \mathbf{a}_n) - \frac{\partial p_i}{\partial \theta}(\boldsymbol{\theta}_n^{**}, \mathbf{a}_0)\right]' n^{1/2}(\boldsymbol{\theta}_n - \boldsymbol{\theta}_0)$$

for $\boldsymbol{\theta}_n^*$, $\boldsymbol{\theta}_n^{**}$ between $\boldsymbol{\theta}_n$ and $\boldsymbol{\theta}_0$. This is $o_p(1)$ whenever the m-vector of derivatives $\partial p_i/\partial\theta$ is continuous and $n^{1/2}(\boldsymbol{\theta}_n - \boldsymbol{\theta}_0)$ converges in distribution. The first two terms in (25) have the form

$$W_n(c_n) - W_n(c)$$

where $c_n \to c(P)$. Now the two general facts of Section 3 apply to convergence in distribution of processes as well. So

$$(W_n, c_n) \xrightarrow{\mathscr{D}} (W_0, c)$$

by the second general fact. The function

$$\varphi(f, t) = f(t) - f(c)$$

is continuous with probability 1 with respect to the distribution of (W_0, c). (Check that: if $f_n \to f$ *uniformly* and $t_n \to c$, then $\varphi(f_n, t_n) \to \varphi(f, c) = 0$. That implies continuity with probability 1 because $W_0(t)$ is continuous, and convergence to a continuous limit function is uniform in the function space at hand.) So by the continuity theorem

$$W_n(c_n) - W_n(c) = \varphi(W_n, c_n) \xrightarrow{\mathscr{D}} \varphi(W_0, c) \equiv 0.$$

Convergence in distribution to a constant is convergence in probability, so we have shown that (25) is $o_p(1)$.

This essentially simple argument can be generalized to multivariate observations and alternative hypotheses. A full treatment appears in [4]. The examples in the next section illustrate the advantages of being free to use data-dependent cells.

7. SOME EXAMPLES

In this section the results of Sections 4 and 6 will be applied to produce several chi-square tests of fit to the family of exponential densities

$$g(x|\theta) = \frac{1}{\theta}e^{-x/\theta} \qquad 0 < x < \infty \tag{26}$$

$$\Omega = \{\theta : 0 < \theta < \infty\}.$$

There are many tests of fit for so standard a family which exceed the chi-square tests in power. Chi-square tests of fit have their greatest potential usefulness in situations where other tests of fit cannot be used (discrete or multivariate data), or where the work of computing critical points for a nontabled distribution is not justified. But restricting ourselves to a single, simple hypothesized family has obvious expository advantages.

Suppose then that $X_1, \ldots, X_n$ are independent random variables having a common unknown distribution which is hypothesized to belong to the family (26).

The Pearson statistic. Choose M fixed cells $E_i = (a_{i-1}, a_i]$ partitioning $(0, \infty)$. The cell probabilities under the null hypothesis are

$$p_i(\theta) = \int_{a_{i-1}}^{a_i} \frac{1}{\theta} e^{-x/\theta}\, dx = a_{i-1}e^{-a_{i-1}/\theta} - a_i e^{-a_i/\theta}.$$

We will estimate θ by the grouped data MLE $\bar{\theta}_n$. The equation resulting from differentiating the logarithm of the multinomial probability function of $N_1, \ldots, N_M$ with respect to θ is

$$\sum_{i=1}^{M} N_i \frac{a_{i-1}e^{-a_{i-1}/\theta} - a_i e^{-a_i/\theta}}{e^{-a_{i-1}/\theta} - e^{-a_i/\theta}} = 0. \tag{27}$$

This has no closed-form solution, but is easily solved iteratively to obtain $\bar{\theta}_n$. Substituting this numerical value into the Pearson statistic

$$\sum_{i=1}^{M} \frac{[N_i - np_i(\bar{\theta}_n)]^2}{np_i(\bar{\theta}_n)}$$

gives a test statistic with approximately the $\chi^2(M-2)$ null distribution.

Using the raw data MLE. The maximum likelihood estimator of θ from $X_1, \ldots, X_n$ is the sample mean, $\hat{\theta}_n = \bar{X}_n$. Who would wish to solve (27) when $\bar{X}_n$ will do our estimating? If $\bar{X}_n$ is used in the Pearson statistic, a θ-dependent limiting null distribution results. But a nice feature of random cells now appears: in testing fit to location-parameter and scale-parameter families, random cells can eliminate the θ-dependence of the null distribution. In this case, we use cells

$$E_i(\bar{X}_n) = (c_{i-1}\bar{X}_n, c_i\bar{X}_n]$$

where the c_i are constants. In the notation of Section 6, $a_{in} = c_i\bar{X}$ and

$$p_i(\theta, \mathbf{a}_n) = \int_{c_{i-1}\bar{X}}^{c_i\bar{X}} \theta^{-1}e^{-x/\theta}\,dx = e^{-c_{i-1}\bar{X}/\theta} - e^{-c_i\bar{X}/\theta}. \tag{28}$$

The estimated cell probabilities $p_i(\bar{X}, \mathbf{a}_n) = e^{-c_{i-1}} - e^{-c_i}$ do not depend on the sample! The Pearson statistic for random cells of this form is algebraically unchanged by the transformation $X_j \to X_j/\theta$ because the cell boundaries move in such a way as to keep the cell frequencies as well as the estimated cell probabilities fixed. Since X_j/θ has the $g(\cdot|1)$ density function when X_j has the $g(\cdot|\theta)$ density function, the distribution of the statistic does not depend on θ. If in particular $c_i = -\log\left(1 - \dfrac{i}{M}\right)$, we obtain M equiprobable cells, $p_i \equiv 1/M$.

The use of random cells thus produces a θ-free null distribution. Not only that, the choice of equiprobable cells simplifies the computation of the statistic and has been shown to have good power properties when fit to a single distribution is being tested. But the limiting distribution is *not* chi-square. It is the distribution of

$$\chi^2(M-2) + \lambda Z_1^2,$$

and requires a special computation to obtain critical points even though λ does not depend on θ.

Using a different quadratic form. Both of the statistics we have thus far applied to testing fit to the family (26) have disabilities in ease of use. The $\mathbf{C}_n$ statistic of Rao and Robson can be explicitly computed, has a $\chi^2(M-1)$ limiting distribution, and also appears from simulations to be more powerful than its two competitors. Let us find its form for this problem. The general forms of both the $\mathbf{C}_n$

and $\mathbf{D}(\hat{\theta}_n)$ statistics from Section 4 simplify because $\sum_1^M \partial p_i/\partial\theta_k = 0$ implies that

$$\mathbf{V}_n'\mathbf{B} = n^{-1/2}\left(\sum_{i=1}^{M} \frac{N_i}{p_i}\frac{\partial p_i}{\partial\theta_1}, \ldots, \sum_{i=1}^{M} \frac{N_i}{p_i}\frac{\partial p_i}{\partial\theta_m}\right).$$

When $m = 1$, the Rao-Robson statistic reduces to

$$R_n = \sum_{i=1}^{M} \frac{(N_i - np_i)^2}{np_i} + \frac{1}{nD}\left(\sum_{i=1}^{M} \frac{N_i}{p_i}\frac{dp_i}{d\theta}\right)^2$$

where

$$D = J - \sum_{i=1}^{M} \frac{1}{p_i}\left(\frac{dp_i}{d\theta}\right)^2$$

and J, p_i, $dp_i/d\theta$ are all evaluated at $\theta = \hat{\theta}_n$.

The use of equiprobable random cells continues to have advantages in simplicity and (probably) in power, so the cells $(c_{i-1}\bar{X}, c_i\bar{X}]$ for $c_i = -\log\left(1 - \frac{i}{M}\right)$ will again be employed. From (28),

$$\left.\frac{dp_i}{d\theta}\right|_{\theta=\bar{X}} = \frac{c_{i-1}\bar{X}}{\theta^2}\, e^{-c_{i-1}\bar{X}/\theta} - \frac{c_i\bar{X}}{\theta^2}\, e^{-c_i\bar{X}/\theta}\Big|_{\theta=\bar{X}}$$

$$= \frac{1}{\bar{X}}(c_{i-1}e^{-c_{i-1}} - c_ie^{-c_i})$$

and a short calculation shows that the test statistic is

$$\frac{M}{n}\sum_{i=1}^{M}\left(N_i - \frac{n}{M}\right)^2 + \frac{M^2}{n}\frac{1}{(1 - M\sum_1^M d_i^2)}\left\{\sum_{i=1}^{M}\left(N_i - \frac{n}{M}\right)d_i\right\}^2.$$

Here $d_i = c_{i-1}e^{-c_{i-1}} - c_ie^{-c_i}$ and N_i is the number of $X_1, \ldots, X_n$ in $(c_{i-1}\bar{X}, c_i\bar{X}]$. This is the recommended chi-square statistic for this problem. The first term is the Pearson statistic for these cells and the raw data MLE, with critical points falling between those of $\chi^2(M - 2)$ and $\chi^2(M - 1)$. The second term need be calculated only if these bounds on the significance of the first term do not yield a sufficiently clear conclusion.

Censored data. It is quite common in experiments on reliability or survival time not to wait for all the lightbulbs to burn out or all the drugged rats to die. The lifetimes are observed in order, so let us denote the ordered observations by

$$X_{(1)} < X_{(2)} < \cdots < X_{(n)}$$

and suppose that observations are stopped after the *sample α-quantile* is observed for some $0 < \alpha < 1$. This is the order statistic $X_{([n\alpha])}$, where $[n\alpha]$ is the greatest integer in $n\alpha$. The exponential distributions form a very common model in life-testing, so it is useful to test fit to this family given only the censored data

$$X_{(1)} < X_{(2)} < \cdots < X_{([n\alpha])}.$$

Here is a challenge to the great flexibility of chi-square methods. The response is to use random cells with boundaries given by sample δ_i-quantiles $\xi_i = X_{([n\delta_i])}$ for

$$0 = \delta_0 < \delta_1 < \cdots < \delta_{M-1} = \alpha < \delta_M = 1,$$

so that the $n - [n\alpha]$ unobserved lifetimes fall in the rightmost cell. Of course, $\xi_0 = 0$ and $\xi_M = \infty$. These random cells fit the demands of Section 6, for the sample quantile ξ_i converges in probability to the population δ_i-quantile $x_i(\theta)$ as $n \to \infty$. The population quantiles are found from

$$\int_0^{x_i} \frac{1}{\theta} e^{-x/\theta}\, dx = \delta_i.$$

This choice of cells produces *nonrandom* cell frequencies

$$N_i = [n\delta_i] - [n\delta_{i-1}],$$

but the theory of Section 6 is entirely undisturbed by this rather odd happenstance. We may cheerfully compute a variety of chi-square statistics for these cells, but we will content ourselves with the Pearson statistic. The grouped data MLE is computed exactly as in (27), "ignoring" the fact that the cell boundaries are random. That is, find $\bar{\theta}_n = \bar{\theta}_n(\xi_1, \ldots, \xi_{M-1})$ numerically by solving

$$\sum_{i=1}^{M} N_i \frac{\xi_{i-1} e^{-\xi_{i-1}/\theta} - \xi_i e^{-\xi_i/\theta}}{e^{-\xi_{i-1}/\theta} - e^{-\xi_i/\theta}} = 0,$$

then use the statistic

$$\sum_{i=1}^{M} \frac{[N_i - np_i(\bar{\theta}_n)]^2}{np_i(\bar{\theta}_n)}$$

with $p_i(\theta) = \xi_{i-1}e^{-\xi_{i-1}/\theta} - \xi_i e^{-\xi_i/\theta}$ and critical points from the $\chi^2(M - 2)$ table. Hats off to the amazing chi-square statistics!

BIBLIOGRAPHY

1. Patrick Billingsley, *Convergence of Probability Measures*, Wiley, New York, 1968.
2. H. Chernoff and E. L. Lehmann, "The use of maximum-likelihood estimates in χ^2 tests for goodness of fit," *Ann. Math. Statist.*, **25** (1954), 579–586.
3. Harold Cramér, *Mathematical Methods of Statistics*, Princeton University Press, Princeton, 1946.
4. D. S. Moore and M. C. Spruill, "Unified large-sample theory of general chi-squared statistics for tests of fit," *Ann. of Statist.*, **3** (1975), 599–616.
5. K. C. Rao and D. S. Robson, "A chi-squared statistic for goodness-of-fit tests within the exponential family," *Comm. in Statist.*, **3** (1974), 1139–1153.

SAMPLE SURVEYS: THEORY AND PRACTICE

Jack E. Graham and J. N. K. Rao

1. INTRODUCTION

The complex structure of all phases of twentieth century society has led to an inevitable ever-increasing need for a myriad of statistics upon which decision making and policy formulation can be based. The sample survey, wherein the characteristics of some population of elements are inferred by observing only a fraction of that population, has made it possible to at least partially answer the demands of both the public and private sectors of our economy for reliable statistics on nearly every area of endeavor. Indeed, the bulk of today's factual information is derived from samples rather than complete enumerations of the populations of interest.

Most responsible citizens have encountered sample surveys in one form or another. National surveys based upon, say, 1500 television viewers are used to reach decisions on what programs are to be cancelled at mid-year. Public opinion polls, which report on the attitudes and preferences of the general population on such diverse topics as political party preferences, feelings about capital punishment, birth control, foreign aid, marriage breakdown and elementary school education, are a common feature of today's newspapers. The great mistrust that many felt in accepting the results of a sample has largely dissipated in light of this increased use of sampling techniques, and the extreme accuracy that many surveys have displayed in, for example, predicting final election results on national television from preliminary returns. With the increased emphasis on statistical thinking at the secondary school level, we can look forward to an even greater appreciation of sampling accompanied by an insistence on sound methodology.

What are the reasons for sampling? Why not conduct a census of all units in the population? Obviously sampling reduces cost as compared to a complete enumeration since only a fraction of the population is observed. Fewer personnel are required and the survey can be completed more quickly. The scope of a sample study can be far broader because the sheer magnitude of a complete census often precludes the possibility of asking detailed and complex questions. The sample, with its limited number of respondents, can often employ a more extensive and probing questionnaire to obtain reliable answers in studies involving complex or sensitive topics, e.g., consumer expenditure patterns or numbers of illegitimate children. Because fewer staff members are required, more rigid standards of excellence may be adhered to in both hiring and training. The sample survey can employ detailed field enumeration procedures which should result in a superior coverage (detection) and enumeration of units. Thus, by reducing nonsampling errors (arising from sources other than the sampling process itself), it is possible that the sample might yield estimates that are in fact more accurate than a complete inquiry might provide. A census might be impractical—the population might be very large or the act of observing destructive: to determine the proportion of defective flashbulbs in a day's production of a factory by firing all of them would be a sterile study! Finally, the sampling approach reduces the response burden on an individual. If all studies were 100%

enumerations, we all would most certainly spend a good portion of each day in supplying responses to one inquiry or another.

But a sample may not be a suitable vehicle for all investigations. A national survey giving estimates with satisfactory precision at the state level would likely provide unacceptable estimates for such smaller areas as the wards of a given city and yet such figures are needed to aid in, say, assigning priorities for senior citizens' homes. One of the most important functions of a census is to provide a current listing (or frame) of units from which future samples may be selected. Hence periodic censuses are essential even in a highly developed country.

The census and sample survey are not always competing alternatives. A census of population or agriculture might collect bench-mark statistics on the age, sex, occupation and marital status of the rural population as well as statistics on farm numbers, crop acreages and livestock numbers. A subset of the population might then be asked to provide additional information on income and expenditures for the twelve month period preceding the enumeration. Surveys are frequently used to provide accurate quarterly estimates for an inter-censal period utilizing census bench-mark data to reduce sampling error. Post-enumeration surveys are now employed to evaluate the quality of the coverage and content of national censuses in a number of countries. Census procedures themselves are often pretested on a relatively small sample of units in a pilot study. The sample survey and census are certainly not strange bedfellows!

In general, two types of surveys may be identified. A descriptive survey concerns itself with the computation of estimates of, say, means and totals for a number of characteristics attached to the units of a population. On the other hand, an analytic survey deals with the relationships existing among variables and involves, for example, the comparison of means and totals for certain subgroups or domains of study of the population, model building and significance testing.

2. BASIC STEPS IN A SAMPLE SURVEY

The success of any sample survey as measured by the quality of the results it ultimately produces is heavily dependent upon the effort and expertise that enters at the planning stage. It is a common misconception that it is a role, indeed a duty, of a statistician to reach into his "bag of tricks" and conjure up a magic formula that will yield, somehow, "good estimates" under the most adverse of circumstances. But in most cases little, if anything, can be done to rescue a survey that has not been carefully thought out in advance. Surveys may differ greatly in their nature, scope and complexity and yet there are a number of steps in conducting a survey that are common to most. There are three major stages: (i) planning, (ii) enumeration, (iii) analysis.

The planning stage itself consists of a number of steps that trace the survey through from its initial conception to the point where the survey is actually undertaken. The steps are not necessarily taken in sequence nor can they always be carried out in a straightforward fashion. They are interrelated and subject to frequent modification. But sound planning is essential if the investigation is to be successful.

The objectives must be laid down by the survey sponsors rather than the survey statisticians who will then assist in formulating the goals in statistical terms. Effective planning is not possible if what the survey is expected to accomplish is not explicitly stated very early in the planning stage.

The study population from which the sample is finally selected may not coincide with the target population for which estimates are required. Thus, geographically isolated farms might be excluded from a survey for they may be inordinately expensive to visit if selected. Because any estimates derived from a survey necessarily refer only to the study population, an assessment must be made of whether the survey objectives will still be met on the basis of supplementary external information about the two populations. It is possible that alternative study populations might be available—should the value of goods shipped be obtained from manufacturers or from truckers? The reliability of the information supplied by the competing study populations and the relative ease of sampling are important factors to be considered in making a choice.

Early in the planning stage decisions must be reached on what data to collect, bearing in mind that the survey objectives must be met but that the respondents must not be overburdened with too many questions. The various totals, means, percentages and other population parameters to be estimated should be explicitly spelled out; any cross-classifications, geographical tabulations or comparisons to be made should be clearly specified.

There may be alternative methods for collecting the required information. Sometimes direct physical observations may be made on units: a psychologist's assistant might himself evaluate the I.Q.'s and reading speeds of a sample of children; objective estimates of potato yields are often made by actually harvesting small plots. Such an approach is costly but it avoids placing undue reliance on a respondent who might, either consciously or unconsciously, provide biased answers. The mail inquiry is a popular means of collecting information because of its relatively low cost. Unfortunately, the response rates to such surveys tend to be low even after repeated reminders are sent out. Marked differences in the characteristics of the response and nonresponse groups often exist so that missing sample members cannot simply be ignored. The mail survey limits the investigator to relatively simple and straightforward questions since the respondent cannot ask for a clarification of points that he might find hazy. The personal interview method in which the enumerator contacts respondents and records the answers to a set of prepared questions is particularly well suited to surveys of a socio-economic nature. The enumerator can solicit the cooperation of the respondent and assist him where there are difficulties in providing an answer. While response rates are improved (but only to the level of 70–75% in typical national or statewide surveys with the use of many call-backs), there is the danger that unless the interviewer is well-trained he may influence the respondent and thus introduce biases into the responses. For example, the question "You go to church regularly, don't you?" would likely draw an affirmative reply irrespective of one's true church-going habits. Such biases, even when averaged over a number of enumerators, can seriously distort the sample results since they often do not tend to balance out. Telephone surveys are sometimes used where feasible as an alternative to personal interviews in order to reduce the costs of travelling and of developing the sampling frame.

A sampling element is the entity on which observations are made. The survey statistician must specify the sampling unit, i.e., the collection of one or more sampling elements, that is most suitable for sampling purposes. For instance, in a consumer expenditure survey one might choose a sample of households (sampling units) and then interview all adults (elements) in each selected household. The sampling units should be well defined and readily recognizable by an enumerator; cost, convenience and statistical efficiency are factors that must be carefully considered.

2.1. The frame. A frame from which sampling units will be selected with known probabilities must be constructed (if not already available). It may be a physical list or perhaps some suitable description of the sampling units in the form of maps or procedures that encompass all units in the study population. It is not essential that the frame should explicitly delineate all units of ultimate interest. In an agricultural study, a primary frame might list counties in some region; after a first-stage sample of counties is selected, a secondary frame might then be constructed which lists for the selected counties either all farms or area segments (nonoverlapping land areas with readily identifiable boundaries).

Because any statistical inference derived from a survey necessarily refers to the universe covered by the frame, it is important that the frame population agree with the study population as much as possible. Frames may be imperfect in a variety of ways. They may be incomplete in that not all units in the study population appear in the frame. Thus, recent arrivals in a city would not appear in a listing of last year's taxpayers. The frame might contain extraneous units that are not members of the study population, e.g., a clothing store might have declared bankruptcy subsequent to the compilation of a list frame of retail outlets. Duplications may occur so that a given unit appears more than once in the frame. The frame could be out-of-date with respect to supplementary information given about the units. The frame might consist of clustered units that have subsequently split. A house which was formerly a single residence might now contain three separate apartments so that three households rather than just one are now defined by the single address given in the frame. The sampler recognizes that in practice perfect frames rarely exist but that frame deficiencies must not be so serious that the survey results will suffer from biases of intolerable magnitudes. Should the effects of the frame problems be small, they might safely be ignored. Otherwise, time and effort must be expended in correcting the frame by detecting and adding missing units, removing duplicates, etc., until the frame approximates the study population (which may itself have to be redefined) to a satisfactory degree. A variety of procedures have been devised to reduce the impact of the deficient frame problems. For example, the multiple frame technique described in Subsection 9.2 permits the use of overlapping frames whose union agrees (within acceptable limits) with the study population.

2.2. The strategy. The survey strategy involves a specification of the selection procedure for bringing a subset of the population elements into the sample (the design) together with formulae for expanding the sample data into estimates of the unknown population parameters, i.e., the estimators. For any given survey a variety of competing strategies will usually be available. The sampling statistician attempts to strike a suitable balance between statistical efficiency and cost. In specifying an efficient design for a given budget, at least rough estimates of various components of variation and of costs appropriate to the survey should be available. Such information might be derived from past surveys or from test studies. Any ancillary information and administrative, physical or operational constraints must also be considered. Numerous compromises will almost certainly be struck before even tentative decisions are reached.

2.3. The questionnaire. Considerable attention must be given to both the type of question asked of the informant and to the sequence in which the questions are asked. The two most common types of forms used to collect information are the questionnaire and the schedule. The questionnaire is a reporting form with specifically worded questions. Either the enumerator or the respondent records the answers. A schedule, rather than incorporating a set of standardized questions, leaves it to the skill and ingenuity of the interviewer to obtain the required information. The questionnaire method has the advantage that the enumerator cannot (at least in theory) bias the answers since the questions are uniformly asked with an exact phrasing. But a rigid questioning structure may not be appropriate for situations where concepts are more complex or conceptually unclear, leading to respondent error. In such cases, a well-trained enumerator using a schedule may be the best means of obtaining accurate information. The two approaches can be combined. The more straightforward information could be left to the individual to fill out and the complicated areas probed by the enumerator.

Questions can be classified as either closed or open. A closed question gives the possible alternative answers as part of the question and the respondent specifies one. An open question, such as "To what magazines do you subscribe?", does not include designated replies. Questions may also be classified as direct or indirect. Thus, "What is your age?" is a direct question, whereas "Give the day, month and year of your birth" indirectly asks the same thing. The latter form might produce a more accurate response, particularly with the elderly who may otherwise be prone to exaggerate their age. In designing the reporting form serious consideration must be given to the question type. Closed questions are more appropriate to mail interview surveys whereas open questions are exceedingly useful in personal interviews where, for instance, attitudes and opinions that may be hard to classify on prior grounds may be requested. The latter has the disadvantage of being more difficult to analyze. Open questions can sometimes be easily converted to closed questions. The question "Is your gross annual income before taxes (a) below $10,000 (b) between $10,000–$20,000 (c) above $20,000?" sacrifices information as compared to "What is your gross annual income before taxes?", but might result in a higher response rate.

There are a few general principles that should be kept in mind in designing a questionnaire. The questions should be clear and as few in number as possible, consistent with the survey objectives. Simple words that convey the exact meaning should be used. Loaded questions, which suggest to the respondent that one particular response is the most desirable, should be avoided. The question "Do you think that the government should introduce tighter gun control legislation so that crimes against society would be reduced?" would be useless to ask since it avoids the arguments given against control and implicitly suggests "Yes" as the appropriate response. The questions appearing in the earlier portion of the form should be straightforward and help to establish rapport with the respondent. Questions that might tend to put the respondent off because of their intimacy, controversial nature, lack of interest to the individual or difficulty to answer should appear towards the end of the form. By this time the interviewee will hopefully be involved with the survey content and sympathetic with the task of the enumerator and the survey goals. The questions should, however, still appear in a logical order, usually progressing from the general to more particular aspects of the survey. There should be a smooth flow from one topic to another. Questions or comments that are not of direct interest to the survey objectives might be inserted in the form to ease the transition.

The questionnaire itself should be attractive looking with adequate spacings between items and a logical physical layout. There should be ample space to record answers. A cluttered page will discourage the best of interviewers or turn off the most willing respondent. The same format should be retained throughout—abrupt changes in layout and style are disturbing and possibly confusing. Symbols can be used to guide the interviewer through complicated pages of the form. Underscoring of certain key words in questions reminds the enumerator to emphasize them in his speech. Arrows and asterisks will lead the enumerator from one part of the form to another in an efficient manner. Boxes can be used in the body of the questionnaire to enclose specific instructions to the individual—capital letters help to draw attention as well. Instructions on the use of the reporting form are given to the interviewer during a training school. But in the case of a form to be filled in by the respondent, explanatory notes and instructions must be provided. Concepts and definitions should be supplied along with typical examples of what to do. In the case of a mail inquiry where an enumerator cannot provide background information, the instructions must be crystal clear. In laying out the questionnaire it should be remembered that the data will ultimately have to be transcribed from the forms by a coder or keypunch operator.

Every effort should be made to expedite the transcription operation, e.g., by providing for the insertion of codes directly on the questionnaire. Certainly good questionnaire design is both an art and a science, coupled with an adequate measure of common sense!

2.4. The pretest. Before a final commitment to a specific survey design is reached, a pretest of the survey should be made. It examines on a small scale the field enumeration procedures and the adequacy of the questionnaire or schedule. Cost components of the field operation can be evaluated. Unforeseen enumeration problems might be detected. For example, refusal to cooperate might prove to be serious and the success of the main survey jeopardized if appropriate actions to alleviate the problem were not devised. Deficiencies in the training program planned for enumerators might become apparent. Should the pretest utilize the same type of sampling units and randomization schemes as the main survey, then estimates of variance components can also be computed. These would be useful in determining, for example, sample sizes and allocations in the main survey.

2.5. The field organization and the survey. With the survey design and reporting form finalized, the day of reckoning is close at hand! The field work for a personal interview survey must be organized and the actual survey conducted. Government agencies which sponsor a number of repeating surveys may have their own permanent field organizations, already staffed with competent enumerators. Less fortunate groups may either set up their own survey staffs or hire an outside survey firm to collect the required information. In all cases enumerators must be appointed and training schools set up. At the training sessions the purposes of the survey are outlined to the prospective field staff by school personnel who have been in close contact with the planning stage of the survey. Interviewing procedures are discussed; the questionnaire or schedule is examined in detail and practice interviews are held both in the classroom and, if possible, with local residents as well. Visual aids such as films may be employed as teaching devices to illustrate successful and poor interviewing techniques. The use of other survey materials such as maps, aerial photographs, dwelling unit listings, visitation records and expense account forms is considered. Emphasis on handling more commonly encountered problems, such as refusal-to-cooperate, not-at-home, unable to find the reporting element, is given. Enumerators are usually advised that any difficulties that they cannot satisfactorily deal with should be referred to their supervisory staff for further action. Enumerators must be briefed on their own specific assignments of elements to be interviewed. Arrangements are made to handle the flow of materials in the field. Completed forms must be periodically collected and reviewed for completeness and glaring inconsistencies by the supervisory staff. Regular enumerator debriefings assist in clearing up many problems provided they are held before specific details of the interviews are forgotten. Otherwise, additional visits might be required to resolve difficulties or to gather further information.

Stage two of the survey—the enumeration—now commences. The headquarters staff usually serve as consultants at this stage, monitoring the progress of the survey or acting in a senior advisor-observer capacity in the field. They must be prepared at all times to modify procedures where necessary to counteract unanticipated problems.

With a mail survey, the questionnaires are dispatched and hopefully returned in short order. Multiple reminders are usually sent out to nonrespondents until a cut-off time is reached. Bias arising from nonresponse can sometimes be reduced by selecting and interviewing a subsample of this group, as will be indicated later.

2.6. Analyzing the survey. The analysis phase of the survey begins when the completed questionnaires are returned from the field. Editing is a diagnostic process used to verify that responses to the various questions on the same form are reasonable, internally consistent and complete. For example, a response of "Four years formal education" cannot be reconciled with an occupation reported as "Doctor." Such nonsampling errors can arise because of an enumerator mistake in wording the question or in recording the answer, the respondent misunderstanding the question, or perhaps even a deliberate attempt to deceive. Missing entries which are not true zero responses are another source of error. Imputation alters one or more of the recorded answers to those questions which fail the edit. The objective is to produce a clean record from which data can be retrieved for analysis. Editing rules must be set down to distinguish acceptable from unacceptable responses. In smaller surveys, editing can be performed by a trained clerical staff working with a manual of editing rules. With larger surveys, hand-editing is often restricted to a more cursory inspection of the record to detect missing entries or glaring errors. The information is, where necessary, then coded, keypunched and transferred to magnetic tape. Computer editing permits the use of complex editing rules which take into account relationships or constraints existing among the many correlated characteristics. As an illustration, it might be known that the following inequalities must be met by two variables: $0.5 \leqslant X/Y \leqslant 4.0$, $X \geqslant 10.3$. The edit fails if either of these pre-identified constraints is not met. A number of such conflicts might be noted on one questionnaire. The entries to be altered must be identified and then corrected on a record by record basis (rather than by adjusting totals). The problem is that one incorrect decision may ultimately cause other previously accepted error checks to become significant and possibly alter the entire character of a questionnaire. One approach employs an algorithm which specifies the minimum number of fields that must be changed to meet all edits and actually identifies these fields. Imputation techniques are then used to estimate rejected field values from unrejected fields in either the same or other records.

Finally, estimates of the characteristics of interest are prepared using the estimation formulae developed at the planning stage. The calculations are usually performed by a computer because of its speed in handling a large data volume and the ease with which complex calculations are carried out. It is now customary to provide customers with an indication of the sampling error to which estimates are subject—often the percentage coefficient of variation, i.e., the ratio of the standard error of the estimate to the estimate itself expressed as a percentage, is supplied. If there are many characteristics, aggregates, means, differences of means, ratios, proportions, etc., it may be impractical to present all sampling errors individually. Sampling errors may then be presented for only the most important items in the survey or for items that exemplify the typical degrees of variation encountered in the survey items. Generalized sampling error tables which relate the magnitude of the estimate to its standard error as an approximation for a set of characteristics are described by Gonzalez *et al.* (1975). Such tables require considerable research and expertise to prepare but they do avoid burdening the reader with a multitude of tables.

The survey report must now be prepared. It can either be a general report for the user who is not primarily interested in the finer details of the survey itself but rather in the published estimates, or it can take the form of a technical report which describes in detail the many steps involved in the survey. The United Nations Statistical Office (1964) has produced an exceedingly useful document on the preparation of sample survey reports. The technical report should be a complete account of all aspects of the survey. It should provide, among other things, details on the design used, the frame, the field enumeration procedures, response rates, the statistical analysis, editing and imputation methods, costs associated with all phases of the survey operation, and so on. In addition, information on various components of variation should be given for a historical record which could be referred to in planning more efficient surveys of a similar nature in the future.

3. A GENERAL SURVEY ERROR MODEL

Consider a finite population U composed of a known number N of distinct units identified through the labels $1, \ldots, N$; the unit receiving label j is denoted by U_j. Suppose that the unknown quantity y_j^* (possibly vector-valued) is associated with U_j, $j = 1, \ldots, N$. Any subset s of U of size $n(s)$ together with the associated observed values is called a sample.[1] Let S denote the set of all possible s and $p(s)$ be the probability with which s is chosen and observed ($p(s) > 0$ and $\sum_{s \in S} p(s) = 1$). The pair (S, p) is called a sample

[1] In some cases it is convenient to define s as an ordered finite sequence $(u_1, \ldots, u_{n(s)})$ where u_j = some U_i and the u_j's need not be distinct. Additional randomizations may also be included in defining s.

design; if $n(s) = n$ for all $s \in S$, then (S, p) is a fixed size design. In practice the listing of all samples S to select an s is not required. Instead, a sampling scheme is used to implement the design.

Sampling theory is largely devoted to the estimation of totals $Y^* = \sum_{i=1}^{N} y_i^*$, means $\bar{Y}^* = Y^*/N$, and ratios $\bar{Y}^*/\bar{X}^*$, where $\bar{X}^*$ is the mean of the values of another character x. However, more and more researchers are becoming interested in the study of complex parameters such as regression and correlation coefficients (see Subsection 9.1).

Suppose that Y^* is to be estimated by observing a sample s with probability $p(s)$. It is assumed that the survey process is conceptually repeatable and that the result of one survey process is independent of any other realization. The response obtained from U_i on trial t, y_{it}, is regarded as a random variable from some hypothetical distribution of responses. Define the indicator random variable a_i as $a_i = 1$ if $U_i \in s$, $a_i = 0$ otherwise and let $y_i = E_t(y_{it} \mid a_i = 1)$, y_i being the expectation of y_{it} over all trials and over all samples containing U_i.

Let $\hat{Y}_t$ be an estimator of Y^* based on the observed y_{it}'s. The total mean square error (MSE) of $\hat{Y}_t$ can be expressed as

$$\begin{aligned} \text{MSE}(\hat{Y}_t) &= E(\hat{Y}_t - Y^*)^2 \\ &= E(\hat{Y}_t - \hat{Y})^2 + E(\hat{Y} - Y)^2 \\ &\quad + 2E[(\hat{Y}_t - \hat{Y})(\hat{Y} - Y)] + (Y - Y^*)^2, \end{aligned} \tag{1}$$

where $\hat{Y}$ is the value of $\hat{Y}_t$ with y_{it} replaced by y_i, E is the expectation taken over all trials and samples and $Y = \sum_1^N y_i$. The first component of (1), $E(\hat{Y}_t - \hat{Y})^2$, is the response variance arising from the variability of the estimator $\hat{Y}_t$ over the trials for a particular s. The second term, $E(\hat{Y} - Y)^2$, is the sampling error component. The third component, $2E[(\hat{Y}_t - \hat{Y})(\hat{Y} - Y)]$, is an interaction due to the relationship between response deviations and sampling errors. The last component is the square of the bias. In view of the relationship (1), the census total MSE could well exceed the total MSE under sampling if the latter employed intensive procedures to reduce response errors. Methods for reducing the sampling error component have been extensively studied in the literature; the emphasis of this paper naturally reflects that concentrated effort. It should not be felt, however, that the assessment and control of response errors and of other sources of nonsampling error are of secondary importance; rather, they are less tractable and hence have received less attention in the past.

4. SAMPLING ERRORS

Suppose that the variate value y_j^* can be known exactly by observing U_j, i.e., $y_{jt} = y_j = y_j^*, j = 1, \ldots, N$. An estimator $\hat{Y}$ is then said to be unbiased for Y if $E(\hat{Y}) = \sum_{s \in S} \hat{Y}_s p(s) \equiv Y$, where $\hat{Y}_s$ is the value of $\hat{Y}$ for the sample s. While unbiasedness is desirable, it is certainly not an essential property of an estimator. A biased estimator might be preferred if it leads to a substantial reduction in the MSE relative to that of an unbiased estimator, provided its bias is small relative to its $\sqrt{\text{MSE}}$.

Most estimators of Y that are considered in the sampling literature are linear in the y_i's and belong to the following general linear class:

$$\hat{Y}_b = \sum_{i \in S} b_{is} y_i, \tag{2}$$

where the coefficients b_{is} are defined for all possible s and all $U_i \in s$. A sample design together with an estimation procedure is termed a strategy. The determination of an optimal strategy for a given budget is a challenging problem facing the survey statistician.

We first outline three basic methods for selecting samples, viz., simple random sampling, systematic sampling and unequal probability sampling. These three methods form the building blocks for the development of a variety of more complex sample designs that are used by survey practitioners.

4.1. Simple random sampling (srs). The set S consists of $\binom{N}{n}$ possible subsets s, each of size n, and $p(s) = 1 \Big/ \binom{N}{n}$ for all s. A commonly used sampling scheme to implement this design is to select the sample unit by unit with equal probability at each draw without replacing the units selected in any of the previous draws. This is accomplished with the aid of a table of random numbers or a computer pseudo-random number generator. The usual unbiased estimator of Y is given by $\hat{Y}_{\text{srs}} = \hat{Y} = N\bar{y}$, where $\bar{y} = \sum_{i\in s} y_i/n$. Note that the choice $b_{is} = b = N/n$ in (2) yields $\hat{Y}$.

Is the estimator $\hat{Y}$ optimal in some statistical sense? This question has received considerable attention in recent years and has led to numerous investigations on such topics as suitable criteria for judging competing estimators in finite populations and the role of maximum likelihood and Bayesian methods of estimation. (See Rao (1976) for a review.) These studies have indicated that $\hat{Y}$ is indeed a "good" estimator of Y when nothing is known about the population frequency distribution of the y-values (e.g., $\hat{Y}$ is a maximum likelihood estimator of Y).

The variance of $\hat{Y}$ is

$$V(\hat{Y}) = \sum_{s\in S} p(s)(\hat{Y}_s - Y)^2 = N^2(N-n)\sigma_y^2/[(N-1)n]$$
$$= N^2\left(1 - \frac{n}{N}\right)\frac{S_y^2}{n}, \tag{3}$$

where $\sigma_y^2 = \sum_1^N (y_i - \bar{Y})^2/N$ is the population variance and $S_y^2 = N\sigma_y^2/(N-1)$ is the population mean square. Formula (3) is easily derived by using the properties of the indicator variable a_i introduced earlier:

$$E(a_i) = \Pr(U_i \in s) = \pi_i = n/N,$$
$$E(a_i a_j) = \Pr(U_i \in s \text{ and } U_j \in s) = \pi_{ij} = n(n-1)/[N(N-1)],$$
$$\hat{Y} = N \sum_{i=1}^{N} a_i y_i/n,$$

where Pr denotes probability. The factor $(N-n)/(N-1)$ is called the finite population correction (or fpc), n/N the sampling rate, and N/n the expansion, jack-up, or blow-up factor.

An unbiased estimator of $V(\hat{Y})$ is given by

$$v(\hat{Y}) = N^2\left(1 - \frac{n}{N}\right)\frac{s_y^2}{n}, \tag{4}$$

where the sample mean square $s_y^2 = \sum_{i\in s}(y_i - \bar{y})^2/(n-1)$ is an unbiased estimator of S_y^2. The standard error (s.e.) of $\hat{Y}$, $\sqrt{v(\hat{Y})}$, estimates the precision of $\hat{Y}$. The coefficient of variation of $\hat{Y}$, $C(\hat{Y}) = \sqrt{V(\hat{Y})}/E(\hat{Y})$, serves as a dimensionless measure of the precision of $\hat{Y}$; $C^2(\hat{Y})$ is known as the rel-variance. The estimated coefficient of variation $c(\hat{Y}) = \sqrt{v(\hat{Y})}/\hat{Y}$ is often reported along with $\hat{Y}$.

It is usually assumed that $\hat{Y}$ is normally distributed for large n so that an approximate 95% confidence interval for Y is given by $\hat{Y} \pm 1.96\sqrt{v(\hat{Y})}$. The normal approximation is justified by appealing to the Central Limit Theorem for finite populations (Madow (1948)). The rate of convergence to normality depends upon n, N and on the shape of the population frequency distribution—in particular, the greater the degree of skewness the slower is the convergence (see Stenlund and Westlund (1975)).

A biased estimator $\tilde{Y}$ will affect the validity of a confidence interval for Y. The degree of distortion in the confidence probability depends upon the bias ratio $|B|/\sqrt{V(\tilde{Y})}$, where $B = E(\tilde{Y}) - Y$ is the bias in $\tilde{Y}$. As a working rule, the effect of bias may be neglected provided that the bias ratio does not exceed 0.1.

It is also of considered practical interest to estimate the proportion P of units belonging to some class. The estimation formulae for proportions readily fall out as a special case of the preceding by setting $y_i = 1$ if U_i possesses the attribute in question, $y_i = 0$ otherwise. Then

$$\hat{P} = \hat{Y}/N = a/n,$$
$$V(\hat{P}) = (N - n)P(1 - P)/[n(N - 1)],$$
$$v(\hat{P}) = (N - n)\hat{P}(1 - \hat{P})/[(n - 1)N],$$

where a is the number of units in the sample belonging to the class.

4.2. Ratio and regression estimation. Suppose that a pair of values (x_i, y_i) is associated with U_i, $i = 1, \ldots, N$, where the y and x characteristics are positively correlated and the population total $X = \sum_{i=1}^{N} x_i$ is known. A sample of n units is selected by SRS and the y and x values are observed. For example, y might be the value of capital construction at a college for the current year and x the corresponding value for the preceding year. If $\bar{y}$ and $\bar{x}$ denote the two sample means, then a ratio estimator of the population total Y is

$$\hat{Y}_r = (\bar{y}/\bar{x})X.$$

Intuitively, $\hat{Y}_r$ recognizes that if $\bar{x}$ is observed to be, say, an underestimate of $\bar{X}$, then $\bar{y}$ will tend to underestimate the unknown $\bar{Y}$ due to the assumed positive correlation between variables. Hence $\bar{y}$ is adjusted to counteract the anticipated underestimation. If interest lies only in the population ratio $R = \bar{Y}/\bar{X}$, then X need not be known; the estimator of R is simply $\hat{R} = \bar{y}/\bar{x}$.

The estimator $\hat{Y}_r$ is not unbiased. Its bias is $B(\hat{Y}_r) = E(\hat{Y}_r) - Y = -\text{Cov}(\hat{Y}_r, \hat{X})/X$, where $\hat{X} = N\bar{x}$. Hence the bias ratio, $|B(\hat{Y}_r)|/\sqrt{V(\hat{Y}_r)} \leqslant \sqrt{V(\hat{X})}/X = C(\hat{X})$ is negligible if $C(\hat{X})$ is small. (Note that $C(\hat{X})$ is of order $1/\sqrt{n}$.) Heuristically, the variance (or MSE) of $\hat{Y}_r$, to order $1/n$, may be derived from the ratio approximation

$$\hat{Y}_r - Y = N(\bar{y} - R\bar{x})\bar{X}/\bar{x} \doteq N(\bar{y} - R\bar{x}).$$

Thus,

$$V(\hat{Y}_r) \doteq N^2\left(\frac{1}{n} - \frac{1}{N}\right)(S_y^2 - 2RS_{yx} + R^2S_x^2), \tag{5}$$

where $S_{yx} = \rho S_y S_x$, ρ = correlation coefficient between y and x, and S_x^2 is the population mean square for x. Empirical investigations have indicated that the approximate variance formula (5) is satisfactory when n is moderately large and $C(\hat{X})$ is small.

Has the use of concomitant information increased precision? A comparison of $V(\hat{Y}_r)$ with $V(\hat{Y})$ shows that $\hat{Y}_r$ is more efficient if $\rho > C_y/(2C_x)$, where $C_y = S_y/\bar{Y}$ and $C_x = S_x/\bar{X}$. In many situations $C_y \doteq C_x$ as in the capital construction example, in which case the condition reduces to $\rho > 0.5$.

An estimator of $V(\hat{Y}_r)$ which might be satisfactory for moderately large n is obtained from (5) by replacing R by $\hat{R}$ and S_y^2, S_x^2, S_{yx} with their corresponding unbiased estimators s_y^2, s_x^2 and s_{yx} where $s_{yx} = \sum_{i \in s}(y_i - \bar{y})(x_i - \bar{x})/(n-1)$ and s_y^2, s_x^2 are the sample mean squares.

Regression estimators of the form

$$\hat{Y}_b = \hat{Y} + b(X - \hat{X}), \tag{6}$$

where $b = s_{yx}/s_x^2$ is the sample regression coefficient, also utilize concomitant x information. The estimator $\hat{Y}_b$ is also biased. For large n the respective variance and estimated variance are approximated by

$$V(\hat{Y}_b) \doteq N^2 S_y^2 (1 - \rho^2)/n, \tag{7}$$

$$v(\hat{Y}_b) \doteq N^2 s_y^2 (1 - r^2)/n, \tag{8}$$

where r is the sample correlation coefficient between y and x. Thus for large n, $V(\hat{Y}_b) < V(\hat{Y})$ unless $\rho = 0$. Also, $V(\hat{Y}_b) < V(\hat{Y}_r)$ if $(\rho\sigma_y - R\sigma_x)^2 > 0$, which is always true unless $\rho\sigma_y/\sigma_x = R$, i.e., unless the true regression line passes through the origin, in which case the approximate variances are equal. However, ratio estimators are usually preferred over regression estimators in practice because of their computational simplicity for more complex sample designs. Limited empirical investigations also indicate that the large-sample variance formulae for $\hat{Y}_r$ are more satisfactory than the corresponding formulae (7) and (8) for $\hat{Y}_b$.

4.3. Systematic sampling. Suppose that the units of a population are serially ordered and that $N = kn$ for some positive integer k. A random integer r in the range 1 to k is selected; the set $s_r = \{U_r, U_{r+k}, U_{r+2k}, \ldots, U_{r+(n-1)k}\}$ is called a systematic sample. The specification of r determines which of the k possible systematic samples will actually be observed. Thus, $p(s_r) = 1/k$ for all r and $\pi_i = 1/k$ for all i, since exactly one of the k possible samples contains U_i.

The usual unbiased estimator of Y is

$$\hat{Y}_{sy} = N \sum_{i \in s_r} y_i/n = N\bar{y}_r, \qquad r = 1, \ldots, k, \tag{9}$$

with variance

$$V(\hat{Y}_{sy}) = N^2 \sum_{r=1}^{k} (\bar{y}_r - \bar{Y})^2/k, \tag{10}$$

$$= N^2 \sigma_y^2 [(1 + (n-1)\rho_w]/n, \tag{11}$$

where σ_y^2 is the population variance and ρ_w, the intraclass correlation coefficient, is a measure of the correlation between pairs of y-values belonging to the same s_r. By comparing (11) with (3), the variance of $\hat{Y}$ under SRS, it is evident that systematic sampling is more efficient if $\rho_w < -1/(N-1)$. Hence the units should be arranged to achieve a maximum within-sample variability. For instance, if auxiliary values x_i are available, then heterogeneous clusters (samples) of units may be induced by listing the units in increasing order of magnitude with respect to these values. Sometimes the ordering occurs naturally; thus systematic sampling has been successfully applied in forest surveys for estimating timber volumes.

If the population units are randomly ordered, then $\rho = -1/(N-1)$ and $V(\hat{Y}_{\text{sy}}) = V(\hat{Y}_{\text{srs}})$. For example, if it may be assumed that units are randomly ordered in a file (or at least approximately so), then systematic sampling may be regarded as a convenient method for selecting a simple random sample of the listed units.

A major disadvantage of systematic sampling is the nonexistence of an unbiased estimator of $V(\hat{Y}_{\text{sy}})$. This is because $\pi_{ij} = 0$ for those U_i and U_j belonging to different systematic samples. (Remark: a necessary and sufficient condition for the existence of an unbiased estimator of a variance (or of any quadratic form) is $\pi_{ij} > 0$ for all $i \neq j = 1, \ldots, N$.)

A number of systematic sampling methods have been proposed in the literature to deal with the general case $N = nk + d$ with $0 < d < k$, the most widely used being the method of circular systematic sampling (Lahiri (1952)). The choice of a random integer r, $1 \leqslant r \leqslant N$, leads to the circular systematic sample consisting of the units U_{r+jk} for $r + jk \leqslant N$, and U_{r+jk-N} for $r + jk > N$, $j = 0, 1, \ldots, n-1$. Unlike the other proposed methods, circular systematic sampling enjoys all three of the following properties: (i) the unbiased estimator of Y is a multiple of the sample mean; (ii) $n(s) = n$ for all samples s; (iii) each s has the same selection probability, viz., $1/N$.

4.4. Unequal probability sampling. Concomitant x information can be incorporated into the design itself by sampling each unit U_i with probability proportional to its size x_i. Such a procedure is intuitively appealing since larger units contribute more to totals than smaller units and, in this sense, are more important. We first consider probability proportional to size sampling with replacement (PPSWR) since the associated theory is simple, unlike that for without replacement designs.

Let p_i denote the selection probability for U_i on a single draw. The selection of n units PPSWR may be accomplished using the method of cumulative totals: (a) without loss in generality assume all x_i's are integers, and form cumulative totals $S_0 = 0$, $S_1 = x_1$, $S_2 = S_1 + x_2, \ldots, S_N = S_{N-1} + x_N = X$; (b) choose a random number R between 1 and X and then select U_i if $S_{i-1} < R \leqslant S_i$. Evidently $\Pr(S_{i-1} < R \leqslant S_i) = p_i = x_i/X$, as required. (c) Repeat step (b) n times, thereby obtaining n units (not necessarily distinct). Lahiri (1951) describes an ingenious alternative technique which avoids the cumulation of sizes: (1) two random numbers are selected, one from 1 to N (say, i) and the other from 1 to x^* (say, k), where $x^* \geqslant \max(x_i)$, $i = 1, \ldots, N$; (2) if $k \leqslant x_i$, then U_i is chosen; otherwise the pair of numbers (i, k) is rejected; (3) steps (1) and (2) are repeated until a unit is selected. The entire procedure is repeated n times to provide a sample of n units. Now the probability that a trial leads to no selection is $Q = 1 - \bar{X}/x^*$. Also, the probability that a trial leads to the selection of U_i is $q_i = x_i/(Nx^*)$. Hence, steps (1) and (2) lead to $p_i = q_i + Qq_i + Q^2q_i + \cdots = x_i/X$. The expected number of trials required for a selection is equal to $x^*/\bar{X}$, which attains its minimum when $x^* = \max x_i$.

The customary unbiased estimator of Y is

$$\hat{Y}_{\text{pps}} = \sum_{i \in s} t_i y_i/(np_i),$$

where $s = (u_1, \ldots, u_n)$, $u_j =$ some U_i, and t_i is the number of times that $U_i \in s$, $\sum_{i=1}^{N} t_i = n$. Because the joint distribution of the t_i is multinomial with probabilities p_i, it readily follows that

$$V(\hat{Y}_{\text{pps}}) = \frac{1}{n}\sum_{1}^{N} p_i\left(\frac{y_i}{p_i} - Y\right)^2. \tag{12}$$

Consequently, PPS sampling will lead to a substantial reduction in the variance as compared to $\hat{Y}_{\text{srs}}$ (in SRS) if y_i is approximately proportional to x_i.

The choice between $\hat{Y}_r$ (in SRS) and $\hat{Y}_{\text{pps}}$ is, however, not as clear cut. If the relationship between y and x is a straight line through the origin and if the conditional variance of y_i about this line for fixed x_i increases faster than x_i, i.e., $V(y_i|x_i) = \delta x_i^g$, $\delta > 0$, $g > 1$, then $V(\hat{Y}_{\text{pps}}) < V(\hat{Y}_r)$. Empirical studies seem to indicate that g usually falls in the range $1 < g < 2$.

An alternative instructive form of (12) is

$$V(\hat{Y}_{\text{pps}}) = \frac{1}{n} \sum_{i<j}^{N}\sum p_i p_j \left(\frac{y_i}{p_i} - \frac{y_j}{p_j}\right)^2. \tag{13}$$

The nonnegative estimator

$$v(\hat{Y}_{\text{pps}}) = \frac{1}{n} \sum_{i<j}^{N}\sum t_i t_j p_i p_j \left(\frac{y_i}{p_i} - \frac{y_j}{p_j}\right)^2 \Big/ E(t_i t_j), \tag{14}$$

suggested by (13) is clearly unbiased for $V(\hat{Y}_{\text{pps}})$. Since $E(t_i t_j) = n(n-1)p_i p_j$, (14) may be recast in the often-quoted form

$$v(\hat{Y}_{\text{pps}}) = \sum_{i \in s} t_i \left(\frac{y_i}{p_i} - \hat{Y}_{\text{pps}}\right)^2 \Big/ [n(n-1)].$$

A disadvantage of PPS sampling is that should the measures of size change with time, $V(\hat{Y}_{\text{pps}})$ will be inflated since the approximate proportionality between y and x will tend to be reduced. Keyfitz (1951) devised a procedure for switching to new probabilities P_i such that the probability of retaining a previously selected U_i is maximized: (i) if $p_i \leqslant P_i$, U_i is retained; (ii) if $p_i > P_i$, then U_i is retained with probability P_i/p_i and rejected with probability $(p_i - P_i)/p_i$; (iii) if U_i is rejected in (ii), then a unit is chosen from among those U_j with $p_j \leqslant P_j$ with probability proportional to $P_j - p_j$. Optimization problems of this nature can be posed as a transportation problem in linear programming (see Raj (1956)).

Despite the fact that the theory for sampling PPSWR is simple, most survey designs employ without replacement (WOR) sampling primarily to avoid the possibility of selecting the same unit more than once. Moreover, the variance, $V(\hat{Y}_{\text{pps}})$, may be further reduced by resorting to a WOR design. A number of WOR strategies have been proposed which may be broadly classified into two Categories: (I) inclusion probability proportional to size (IPPS) designs (i.e., $\pi_i \propto x_i$) with the estimator $\hat{Y}_{\text{HT}} = \sum_{i \in s} y_i/\pi_i$ (Horvitz-Thompson (1952)); (II) non-IPPS designs using estimators other than $\hat{Y}_{\text{HT}}$ for which the variance becomes zero when $y_i \propto x_i$ as in the case of (I). Attention to the following considerations was given in developing these methods: (a) the variance of the estimator should be smaller than $V(\hat{Y}_{\text{pps}})$; (b) the unbiased variance estimator should be non-negative; (c) the computations involved should be relatively simple. The theory underlying such methods is mathematically challenging and has attracted considerable attention in recent years.

By writing $\hat{Y}_{\text{HT}} = \sum_1^N a_i y_i/\pi_i$, where a_i is the indicator random variable, and noting that $\sum_{j \neq i} \pi_{ij} = (n - 1)\pi_i$ for any fixed size design, it may be verified that

$$V(\hat{Y}_{\text{HT}}) = \sum_{i<j}^{N}\sum (\pi_i\pi_j - \pi_{ij})\{(y_i/\pi_i) - (y_j/\pi_j)\}^2.$$

This form of $V(\hat{Y}_{\text{HT}})$ leads to the unbiased variance estimator

$$v(\hat{Y}_{\text{HT}}) = \sum_{\substack{i<j \\ \in s}}\sum (\pi_i\pi_j - \pi_{ij})\{(y_i/\pi_i) - (y_j/\pi_j)\}^2/\pi_{ij},$$

known as the "Sen-Yates-Grundy estimator."

In large-scale surveys the population is sometimes subdivided (stratified) to the extent that $n = 2$ clusters of units (primary sampling units or psu's) are selected with unequal probabilities WOR from each stratum. Among the Category I methods for $n = 2$, we describe that due to Brewer (1963): the first unit is selected with "working probabilities" p_i^* and the second unit with probabilities proportional to the size measures x_i of the remaining population units. Thus,

$$\pi_i = p_i^* + \sum_{j \neq i}^{N} p_j^* p_i/(1 - p_j),$$

and the solution for $\pi_i = 2p_i$ is

$$p_i^* = \frac{2p_i(1 - p_i)/(1 - 2p_i)}{1 + \sum_{j=1}^N [p_j/(1 - 2p_j)]}.$$

The joint inclusion probability π_{ij} is given by

$$\pi_{ij} = \frac{p_i^* p_j}{(1 - p_i)} + \frac{p_j^* p_i}{(1 - p_j)}.$$

For the case $n > 2$ in Category I, Sampford's (1967) method appears to be the most attractive. The steps in his rejective sampling scheme are: (i) select the first unit with PPS; (ii) make subsequent selections WR sequentially with probabilities proportional to $p_i/(1 - np_i)$; (iii) accept only those selections that give a unit different from those previously drawn; (iv) continue until n distinct units are selected. The theory underlying Sampford's method is somewhat complex but the computation of the π_{ij}'s is not cumbersome for small n, e.g., $n < 10$. For $n = 2$ the methods of Brewer and Sampford are equivalent, i.e., lead to the same design $p(s)$.

Turning now to Category II, we present two methods: (i) Murthy's (1957) strategy; (ii) the random group method of Rao-Hartley-Cochran (1962). Murthy's method for $n = 2$ is as follows: the first unit is selected with PPS and the second unit with PPS of the remaining units (i.e., PPSWOR). Murthy's unbiased estimator of Y is given by

$$\hat{Y}_{\mathrm{M}} = \frac{(1 - p_j)(y_i/p_i) + (1 - p_i)(y_j/p_j)}{2 - p_i - p_j}, \qquad s = \{i, j\}.$$

Noting that

$$p(s) = E(a_i a_j) = \frac{p_i p_j}{1 - p_i} + \frac{p_j p_i}{1 - p_j},$$

it may be verified that

$$V(\hat{Y}_{\mathrm{M}}) = \sum_{i<j}^{N}\sum p_i p_j \frac{1 - p_i - p_j}{2 - p_i - p_j}\left(\frac{y_i}{p_i} - \frac{y_j}{p_j}\right)^2,$$

with the unbiased variance estimator

$$v(\hat{Y}_{\mathrm{M}}) = \frac{(1 - p_i - p_j)(1 - p_i)(1 - p_j)}{(2 - p_i - p_j)^2}\left(\frac{y_i}{p_i} - \frac{y_j}{p_j}\right)^2.$$

Murthy's method is applicable for general n but the computations become complex as n increases. However, the unbiased variance estimator for the general case is nonnegative.

The random group method remains simple for any n, unlike the other proposed techniques in Categories I and II. A sample s is obtained in two steps: (i) the population is divided at random into n groups $G_1, \ldots, G_n$ of fixed sizes $N_1, \ldots, N_n$ respectively, $\sum_{l=1}^{n} N_l = N$; (ii) one unit is then selected independently from each random group with selection probabilities $\tilde{p}_j = p_j/P_l$ for G_l, where $P_l = \sum_{t\in G_l} p_t$. The unbiased estimator of Y proposed by Rao-Hartley-Cochran is

$$\hat{Y}_{\text{RHC}} = \sum_{i=1}^{n} \frac{y_i}{\tilde{p}_i},$$

where the suffix i denotes the unit selected from group i and $\tilde{p}_i$ depends on s.

For a given split of the population, $\hat{Y}_{\text{RHC}}$ is conditionally unbiased, i.e., $E_2(\hat{Y}_{\text{RHC}}) = \sum_1^n Y_i = Y$ (where the subscript 2 denotes "for a given split"), with conditional variance

$$V_2(\hat{Y}_{\text{RHC}}) = \sum_{i=1}^{n} V_2(y_i/\tilde{p}_i).$$

Here

$$\begin{aligned} V_2(y_i/\tilde{p}_i) &= \sum_{\substack{t<t' \\ \in G_i}}\sum \tilde{p}_t\tilde{p}_{t'}\{(y_t/\tilde{p}_t) - (y_{t'}/\tilde{p}_{t'})\}^2 \\ &= \sum_{\substack{t<t' \\ \in G_i}}\sum p_t p_{t'}\{(y_t/p_t) - (y_{t'}/p_{t'})\}^2, \end{aligned}$$

where $Y_i = \sum_{t\in G_i} y_t$ (see (13) with $n = 1$). Consequently, the overall variance of $\hat{Y}_{\text{RHC}}$ is given by

$$\begin{aligned} V(\hat{Y}_{\text{RHC}}) &= EV_2(\hat{Y}_{\text{RHC}}) + VE_2(\hat{Y}_{\text{RHC}}) = EV_2(\hat{Y}_{\text{RHC}}) \\ &= \sum_{i=1}^{n} EV_2(y_i/\tilde{p}_i). \end{aligned}$$

Since $[N_i(N_i - 1)]/[N(N - 1)]$ is the probability that in a random split a pair of units falls into G_i, we have

$$EV_2(y_i/\tilde{p}_i) = \frac{N_i(N_i - 1)}{N(N - 1)} \sum_{t<t'}^{N}\sum p_t p_{t'}\{(y_t/p_t) - (y_{t'}/p_{t'})\}^2.$$

Thus,

$$V(\hat{Y}_{\text{RHC}}) = \frac{\sum_{i=1}^{n} N_i^2 - N}{N(N - 1)} \sum_{1}^{N} p_t\left(\frac{y_t}{p_t} - Y\right)^2. \tag{15}$$

The optimal choice of the N_i's which minimizes (15) is given by $N_1 = \cdots = N_k = R + 1$ and $N_{k+1} = \cdots = N_n = R$, where $N = nR + k$ for some positive integer R, $(0 \leqslant k < n)$. If $N = nR$, (15) reduces to

$$V(\hat{Y}_{\text{RHC}}) = \left(1 - \frac{n-1}{N-1}\right) V(\hat{Y}_{\text{pps}})$$

under the optimal choice.

An unbiased estimator of $V(\hat{Y}_{\text{RHC}})$ is

$$v(\hat{Y}_{\text{RHC}}) = \frac{\sum_{i=1}^{n} N_i^2 - N}{N^2 - \sum_{i=1}^{n} N_i^2} \sum_{i=1}^{n} P_i \left(\frac{y_i}{p_i} - \hat{Y}_{\text{RHC}}\right)^2.$$

The random group method is ideally suited for changing to new size measures since the Keyfitz scheme can be applied to each group independently. Moreover, supplementary studies involving a reduced sample size can be accommodated using one (or more) of the n groups since each group provides an unbiased estimator of the population total.

Extensive empirical and semi-empirical results on the relative efficiencies of the estimators and variance estimators in Categories I and II are given by Rao and Bayless (1969) for $n = 2$ and by Bayless and Rao (1970) for $n = 3, 4$. Work on the Madow-type central limit theorems for the estimators in unequal probability sampling has also been reported (see, e.g., Rosén (1974)).

We note that $\hat{Y}_{\text{HT}}$, $\hat{Y}_{\text{M}}$ and $\hat{Y}_{\text{RHC}}$ all belong to the linear class (2): $\hat{Y}_{\text{HT}}$ uses $b_{is} = b_i = 1/(np_i)$; $\hat{Y}_{\text{M}}$ uses $b_{is} = (1 - p_j)/[p_i(2 - p_i - p_j)]$; $\hat{Y}_{\text{RHC}}$ uses $b_{is} = 1/\tilde{p}_i$.

5. COMMONLY USED SURVEY STRATEGIES

Subsections 5.1 to 5.6 describe some of the more commonly used strategies in survey sampling.

5.1. Unistage cluster sampling. The population consists of (or is divided into) M clusters (groups of units), the size of the ith cluster being M_i, $\sum_{i=1}^{M} M_i = N$. A sample of m clusters is selected by any one of the three basic methods already outlined and all units in the sampled clusters are observed. For example, in a survey dealing with the reading habits of freshmen in an introductory English course having M sections, one could select a sample of m sections and interview all students in these sections. Evidently cluster sampling in this case is operationally easier, faster, and hence cheaper than selecting a sample of students directly from the N students in the population. Sometimes a frame of the population units may not be available in which case some form of cluster sampling becomes a necessity. Thus, in an urban family income survey, a satisfactory frame of individual households might not be available or very costly to compile. A sample of city blocks could, however, be selected from a listing of all city blocks (e.g., from a map) and all households in the selected blocks visited.

Let $Y_i = \sum_{j=1}^{M_i} y_{ij}$ and $\bar{Y}_i = Y_i/M_i$ denote, respectively, the population total and mean for the ith cluster, where y_{ij} is the variate value for the jth unit in the ith cluster. The relevant formulae for estimating the total $Y = \sum Y_i$ can be readily obtained from those for the three basic methods already outlined. Thus, letting s_1 denote a sample of m clusters, we have

$$\hat{Y}_{\text{srs}} = \frac{M}{m} \sum_{i \in s_1} Y_i,$$

$$\hat{Y}_r = \left(\sum_{i \in s_1} Y_i \Big/ \sum_{i \in s_1} X_i \right) X,$$

where X_i is the ith cluster total for the concomitant variable x. Similarly

$$\hat{Y}_{\text{pps}} = \sum_{i \in s_1} t_i Y_i/(np_i),$$

where $p_i = X_i/X$; the choice $X_i = M_i$ is commonly used. The formulae for the variance (MSE) are similarly found, e.g.,

$$V(\hat{Y}_{\text{pps}}) = \frac{1}{m} \sum_{i=1}^{M} p_i \left(\frac{Y_i}{p_i} - Y \right)^2,$$

$$v(\hat{Y}_{\text{pps}}) = \frac{1}{m(m-1)} \sum_{i \in s_1} t_i \left(\frac{Y_i}{p_i} - \hat{Y}_{\text{pps}} \right)^2.$$

In the special case of equi-sized clusters, i.e., $M_i = \bar{M}$, the variance formula for $\hat{Y}_{\text{srs}}$ may be recast for large N as

$$V(\hat{Y}_{\text{srs}}) \doteq N^2 \left(\frac{1}{m} - \frac{1}{M} \right) \frac{S^2}{\bar{M}} [1 + (\bar{M} - 1)\rho_c], \tag{16}$$

where $S^2 = \sum \sum (y_{ij} - \bar{Y})^2/(N-1)$, $\bar{Y} = Y/N$, and where ρ_c is the intraclass correlation coefficient between pairs of observations within the same cluster:

$$\rho_c = \frac{2\sum_{i}^{M}\sum_{j<j'}^{\overline{M}}\sum (y_{ij} - \overline{Y})(y_{ij'} - \overline{Y})}{(\overline{M} - 1)(N - 1)S^2}.$$

In most cluster sampling situations ρ_c is positive since units within the same cluster tend to resemble each other to a greater degree than those belonging to different clusters.

Comparing (16) with the corresponding formula (3) for a simple random sample of $n = m\overline{M}$ units, it is seen that cluster sampling is less efficient if $\rho_c > 0$. However, because the cost per unit is often appreciably lower when units are picked up in clusters, the precision achieved under cluster sampling for a given cost may exceed that realized under SRS of the units.

5.2. Multistage designs. If the clusters are large and relatively homogeneous, it may be more efficient (for the same cost) to subsample clusters rather than to survey all units in the selected clusters. In the reading habits example, one could select more classes without incurring additional cost if the students in these classes were subsampled. The theory for subsampling (or multistage designs) can be readily developed through the repeated use of conditional expectation and conditional variance operators. We outline the theory for the general linear class of estimators:

$$\hat{\hat{Y}}_b = \sum_{i \in s_1} b_{is_1} \hat{Y}_i,$$

where $\hat{Y}_i$ is an unbiased estimator of the ith selected cluster total based on the second and lower stages of sampling (if any), i.e., $E_2 \hat{Y}_i = Y_i$, where E_2 denotes conditional expectation for a given sample s_1 of clusters selected at the first stage. Subsampling is performed independently within each of the selected clusters.

The variance of $\hat{\hat{Y}}_b$ may be written as the sum of two components representing between and within cluster variabilities. Thus

$$\begin{aligned} V(\hat{\hat{Y}}_b) &= VE_2(\hat{\hat{Y}}_b) + EV_2(\hat{\hat{Y}}_b) \\ &= V(\hat{Y}_b) + E\left(\sum_{i \in s_1} b_{is_1}^2 V_{is_1}\right), \end{aligned} \tag{17}$$

where $\hat{Y}_b = \sum_{i \in s_1} b_{is_1} Y_i$, $V_{is_1} = V_2(\hat{Y}_i)$, and V_2 is the conditional variance operator for a given s_1. In most applications V_{is_1} is independent of s_1, i.e., $V_{is_1} = V_i$, so our discussion will be confined to this special case. The second term of (17) (the within cluster component) then reduces to $\sum_{i=1}^{M} E(\tilde{b}_{is_1}^2)V_i$, where $\tilde{b}_{is_1} = b_{is_1}$ if $i \in s_1$, $\tilde{b}_{is_1} = 0$ otherwise. The first term of (17), $V(\sum b_{is_1} Y_i)$, is the variance obtained under unistage cluster sampling. Through the repeated use of (17), a formula for the conditional variance V_i can be explicitly developed.

Suppose
$$f(\mathbf{Y}) = \sum d_{is_1} Y_i^2 + \sum_{i<j}\sum d_{ijs_1} Y_i Y_j$$
is an unbiased variance estimator under unistage cluster sampling, where d_{is_1} and d_{ijs_1} are prespecified constants (like b_{is_1}). Then, noting that $E(\tilde{d}_{is1}) = V(\tilde{b}_{is_1})$, where $\tilde{d}_{is_1} = d_{is_1}$ if $i \in s_1$, $\tilde{d}_{is_1} = 0$ otherwise, an unbiased estimator of $V(\hat{\hat{Y}}_b)$ is obtained as

$$v(\hat{\hat{Y}}_b) = f(\hat{\mathbf{Y}}) + \sum b_{is_1} \hat{V}_i, \tag{18}$$

where $\hat{V}_i$ is an unbiased estimator of V_i based on the subsampled units. This immediately leads to the following convenient rule for unbiased variance estimation (Raj (1966)): (i) obtain a copy $f(\hat{\mathbf{Y}})$ of the unistage variance estimator $f(\mathbf{Y})$ by replacing Y_i by $\hat{Y}_i$; (ii) also obtain a copy of $\hat{Y}_b$ by replacing Y_i with $\hat{V}_i$; (iii) the sum of these two copies is the desired variance estimator. An explicit formula for $\hat{V}_i$ can be derived through repeated use of (18).

The foregoing general theory will now be illustrated for a two-stage design. Suppose the first-stage units (clusters) are sampled with IPPS of the clusters and m_i units are subsampled from cluster i by SRS whenever $i \in s_1$. If the Horvitz-Thompson estimator, $\hat{\hat{Y}}_b = \sum \hat{Y}_i/\pi_i$, is employed and $f(\mathbf{Y})$ is the Sen-Yates-Grundy variance estimator, then

$$b_{is_1} = 1/\pi_i = N/(mM_i) = E(\tilde{b}_{is_1}^2), \qquad \hat{Y}_i = M_i\bar{y}_i,$$
$$V_i = M_i^2\{(1/m_i) - (1/M_i)\}S_i^2, \qquad \hat{V}_i = M_i^2\{(1/m_i) - (1/M_i)\}s_i^2,$$

$$f(\hat{\mathbf{Y}}) = \sum_{\substack{i<j \\ \in s_1}}\sum \frac{\pi_i\pi_j - \pi_{ij}}{\pi_{ij}} \left(\frac{\hat{Y}_i}{\pi_i} - \frac{\hat{Y}_j}{\pi_j}\right)^2,$$

$$V(\hat{Y}_b) = \sum_{i<j}\sum (\pi_i\pi_j - \pi_{ij})\left(\frac{Y_i}{\pi_i} - \frac{Y_j}{\pi_j}\right)^2,$$

where $\bar{y}_i$, S_i^2 and s_i^2 are, respectively, the sample mean, the population mean square and the sample mean square corresponding to the ith cluster, and π_{ij} is the joint inclusion probability for clusters i and j. When $m_i \equiv c$, $\hat{\hat{Y}}_b$ reduces to a multiple of the sample total, i.e., $\hat{\hat{Y}}_b$ is a self-weighting estimator.

The extension of this theory to commonly used ratio estimators of the form $(\hat{\hat{Y}}_b/\hat{\hat{X}}_b)X$ is straightforward.

Unbiased variance estimation in multistage designs is greatly simplified if psu's are sampled with replacement and subsampling is performed independently each time a psu is selected. In the case of PPS sampling with replacement, an unbiased estimator of Y is

$$\hat{\hat{Y}}_{\text{pps}} = \sum_{i=1}^{m} \hat{Y}_i'/(mp_i'),$$

where $\hat{Y}_i'$ is an unbiased estimator of the total for the psu chosen at draw i, $i = 1, \ldots, m$, and p_i' is the corresponding selection probability. Because the random variables $\hat{Y}_i'/p_i'$, $i = 1, \ldots, m$, are stochastically independent unbiased estimators of Y, it immediately follows that an unbiased variance estimator is given by the simple formula

$$v(\hat{\hat{Y}}_{\text{pps}}) = \frac{1}{m(m-1)} \sum_{i=1}^{m} \left(\frac{\hat{Y}_i'}{p_i'} - \hat{\hat{Y}}_{\text{pps}} \right)^2.$$

Note that primaries may be sampled under any design whatsoever as long as the Y_i's can be unbiasedly estimated. No such simple formula for the variance $V(\hat{\hat{Y}}_{\text{pps}})$ which depends only on the variation among psu totals can be obtained. For instance, in a two-stage design involving SRSWOR of secondaries, the variance of $\hat{\hat{Y}}_{\text{pps}}$ is given by

$$V(\hat{\hat{Y}}_{\text{pps}}) = \frac{1}{m} \sum_{1}^{M} p_i \left(\frac{Y_i}{p_i} - Y \right)^2 + \frac{1}{m} \sum_{1}^{M} \frac{M_i^2}{p_i} \left(\frac{1}{m_i} - \frac{1}{M_i} \right) S_i^2. \tag{19}$$

5.3. Double sampling. In the discussion of ratio and regression estimation for unistage sampling (Subsection 4.2), we assumed that the population total X of the auxiliary variable x was known. If such information is lacking, and the cost per unit of observing x alone is much less than for the characteristic of interest y, it may be more efficient (for the same cost) to adopt double sampling (two-phase sampling): a large sample s' of size n' is selected at the first phase and x alone is observed; at the second phase a subsample $\tilde{s}$ of size n is drawn from s' and y is observed.

If SRS is adopted at both phases, a ratio estimator of Y, corresponding to $\hat{Y}_r$, is

$$\hat{Y}_{rd} = (\bar{y}/\bar{x})\hat{X}',$$

where $\bar{y}$ and $\bar{x}$ are the means from the subsample $\tilde{s}$, $\hat{X}' = N\bar{x}'$ is the estimator of X calculated from the first-phase sample and $\bar{x}' = \sum_{i\in s'} x_i/n'$.

As in the case of $\hat{Y}_r$, the variance (MSE) of $\hat{Y}_{rd}$ for large n may be derived heuristically from the approximation

$$\begin{aligned}\hat{Y}_{rd} - Y &= N\left[\frac{\bar{X}}{\bar{x}}(\bar{y} - R\bar{x}) + \frac{\bar{y}}{\bar{x}}(\bar{x}' - \bar{X})\right]\\ &\doteq N[(\bar{y} - R\bar{x}) + R(\bar{x}' - \bar{X})].\end{aligned}$$

Noting that

$$\begin{aligned}E_2(\hat{Y}_{rd} - Y) &\doteq N(\bar{y}' - \bar{Y}),\\ V_2(\hat{Y}_{rd} - Y) &\doteq N^2V_2(\bar{y} - R\bar{x}),\end{aligned}$$

where $\bar{y}' = \sum_{i\in s'} y_i/n'$ and E_2 and V_2 are the conditional expectation and conditional variance operators for a given s', we get

$$V(\hat{Y}_{rd}) = V(\hat{Y}_{rd} - Y) \doteq V(\hat{Y}_r) + \frac{N^2}{n'}\left(1 - \frac{n'}{N}\right)(2RS_{yx} - R^2S_x^2), \tag{20}$$

with $V(\hat{Y}_r)$ given by (5). Formula (20) is correct to terms of order $1/n$. A consistent estimator of $V(\hat{Y}_{rd})$ can be obtained from (20), as in the case of $V(\hat{Y}_r)$, by replacing the unknown parameters by their estimators.

The extension of double sampling to multiphase sampling involving more than two phases is straightforward. In a multistage cluster design, one could employ multiphase sampling at any of the stages.

5.4. Stratified sampling. With a heterogeneous population of units, significant increases in the precision of estimates of population characteristics may often be achieved by partitioning the first-stage sampling units (psu's) into relatively homogeneous groups (strata) using either qualitative or quantitative auxiliary information. Independent samples of designated sizes are then drawn according to some specified design from each stratum. Stratification may also be utilized within selected psu's if warranted but the discussion here is confined to the case where only psu's are stratified.

Aside from the worthwhile reductions in variance that usually result, other advantages of stratification are: (i) separate estimates for individual strata (or stratum groups) may be obtained whenever needed; (ii) increased flexibility: alternative sampling strategies can be used within different strata; (iii) administrative convenience.

As an example, Bayless, Shah and Finkner (1973) report on an educational assessment survey conducted in the State of Maine to provide information on achievements in reading, mathematics, writing, etc., relative to important demographic, school and home background variables. Public and nonpublic schools were divided into 37 strata based upon geographic region, size of school, and the per pupil expenditure of the school. A sample of 97 schools participated in the study; from each school a random sample of twenty 17-year-old students was selected (except for 14 large schools from which 40 pupils were selected).

The estimation theory for totals (or means) with a stratified design is a straightforward extension of the results already presented. Suppose that Y_h denotes the total for stratum h, $h = 1, \ldots, L$, so that $Y = Y_1 + \cdots + Y_L$. Let $\hat{Y}_h$ denote an unbiased estimator of Y_h based on the sample from stratum h. Then an unbiased estimator of Y is given by $\hat{Y}_{\text{st}} = \sum_{h=1}^{L} \hat{Y}_h$. Noting that the $\hat{Y}_h$'s are independent, we have the variance $V(\hat{Y}_{\text{st}}) = \sum V(\hat{Y}_h)$ and the variance estimator $v(\hat{Y}_{\text{st}}) = \sum v(\hat{Y}_h)$, where $V(\hat{Y}_h)$ and $v(\hat{Y}_h)$ denote the variance and estimated variance of $\hat{Y}_h$ respectively.

In the case of ratio estimation, two different estimators of Y are available: a separate ratio estimator $\hat{Y}_{\text{rs}} = \sum_h \hat{Y}_{\text{rh}} = \sum (\hat{Y}_h/\hat{X}_h) X_h = \sum \hat{R}_h X_h$ and a combined ratio estimator $\hat{Y}_{\text{rc}} = (\hat{Y}_{\text{st}}/\hat{X}_{\text{st}})X$, where X_h is the hth stratum total for the concomitant variable x, $\hat{X}_h$ is the corresponding estimator, and $X = \sum X_h$. Note that $\hat{Y}_{rs}$ requires a knowledge of the individual totals X_h, unlike $\hat{Y}_{rc}$. If m_h, the number of psu's selected from stratum h, is sufficiently large for all h, then the ratio approximation applied to each stratum leads to $V(\hat{Y}_{rs}) \doteq \sum V(\hat{Y}_h - R_h\hat{X}_h)$. The variance estimator is obtained from $V(\hat{Y}_{rs})$ by substituting estimates for the unknown parameters, i.e., $v(\hat{Y}_{rs}) \doteq \sum [v(\hat{Y}_h) + \hat{R}_h^2 v(\hat{X}_h) - 2\hat{R}_h c(\hat{Y}_h, \hat{X}_h)]$, where c is the sample covariance. On the other hand, the ratio approximation applied to $\hat{Y}_{rc}$ is valid provided only that $m = \sum m_h$ is large. Thus, $V(\hat{Y}_{rc}) \doteq V(\hat{Y}_{st} - R\hat{X}_{st})$, and $v(\hat{Y}_{rc}) \doteq \sum [v(\hat{Y}_h) + \hat{R}_{st}^2 v(\hat{X}_h) - 2\hat{R}_{st} c(\hat{Y}_h, \hat{X}_h)]$, where $\hat{R}_{st} = \hat{Y}_{st}/\hat{X}_{st}$.

As an illustration, consider a stratified sample $s = (s_1, \ldots, s_L)$ where s_h is a simple random sample of n_h units from the N_h units in stratum h. Then

$$\hat{Y}_{st} = \sum N_h \bar{y}_h, \qquad \hat{X}_{st} = \sum N_h \bar{x}_h,$$

$$V(\hat{Y}_{st}) = \sum N_h^2\{(1/n_h) - (1/N_h)\}S_{yh}^2,$$

$$V(\hat{Y}_{rs}) \doteq \sum N_h^2\{(1/n_h) - (1/N_h)\}(S_{yh}^2 - 2R_h S_{yxh} + R_h^2 S_{xh}^2),$$

$$V(\hat{Y}_{rc}) \doteq \sum N_h^2\{(1/n_h) - (1/N_h)\}(S_{yh}^2 - 2RS_{yxh} + R^2 S_{xh}^2).$$

A comparison of $V(\hat{Y}_{st})$ with $V(\hat{Y}_{srs})$ shows that stratification will result in reduced variance if n_h is proportional to N_h (proportional allocation) or to $N_h S_h$ (Neyman allocation); see Subsection 7.2.

If the relationship between y and x is a straight line through the origin in every stratum, then $R_h = S_{yxh}/S_{xh}^2$ and

$$V(\hat{Y}_{rc}) - V(\hat{Y}_{rs}) \doteq \sum N_h^2\{(1/n_h) - (1/N_h)\}(R - R_h)^2 S_{xh}^2.$$

Consequently, $\hat{Y}_{rs}$ will lead to significant gains in efficiency if the stratum ratios differ appreciably from one another, provided n_h is large for all h. On the other hand, if the number of strata is large and the n_h's are small, the absolute bias in $\hat{Y}_{rs}$ may not be negligible relative to its standard error, unlike $\hat{Y}_{rc}$.

5.5. Post-stratification. If samples cannot be drawn separately from the strata (e.g., if the individual stratum frames are not available), but the stratum sizes are known (at least approximately), then a sample selected from the overall population can be post-stratified by assigning each unit to its appropriate stratum after selection. The post-stratified estimator of Y in SRS, for instance, is $\hat{Y}_{st(p)} = \sum N_h \bar{y}_h$, where $\bar{y}_h$ is the sample mean of the units falling into stratum h. The sample size n_h is now a random variable and we assume that $n_h > 0$ (with probability one) for all h. Since for given $n_1, \ldots, n_L$ we have stratified SRS, the variance of $\hat{Y}_{st(p)}$ reduces to

$$V(\hat{Y}_{st(p)}) \doteq E\left[\sum N_h^2\{(1/n_h) - (1/N_h)\}S_h^2\right] = [(N/n) - 1]\sum N_h S_h^2,$$

using the approximation $E(1/n_h) = 1/E(n_h)$. If the sampling fraction n/N is negligible, then $V(\hat{Y}_{st(p)}) < V(\hat{Y}_{srs})$. Under proportional allocation, $n_h = n(N_h/N)$, the variance of $\hat{Y}_{st}$ is identical with $V(\hat{Y}_{st(p)})$ as given above.

5.6. Double sampling for stratification. Suppose we wish to stratify a population of size N according to the values of an auxiliary variable x, but the frequency distribution of x is unknown. A large sample of size n' is taken by, say, SRS and the x-values are observed; these selected units are then classified into L strata of size n'_h, $\sum n'_h = n'$, on the basis of these values. Independent subsamples of sizes $n_h = v_h n'_h$, where v_h is a constant such that $0 < v_h \leqslant 1$, are selected by SRS from the L strata, and the character of interest, y, is observed. If we assume that n' is so large that $n'_h > 0$ with probability one for all h, then an unbiased estimator of Y is $\hat{Y}_d = N \sum n'_h \bar{y}_h / n'$, where $\bar{y}_h$ is the mean of the n_h units in stratum h. The variance of $\hat{Y}_d$ is then given by

$$V(\hat{Y}_d) = N^2\left(\frac{1}{n'} - \frac{1}{N}\right)S^2 + N^2 \sum \frac{W_h S_h^2}{n'}\left(\frac{1}{v_h} - 1\right), \tag{21}$$

where S^2 is the population mean square, S_h^2 is the mean square in stratum h and W_h is the true weight of stratum h. This double sampling procedure is viable only if the cost of identifying the first-phase sample members is small relative to the cost of collecting the second-phase y-values. For example, it would be worthwhile when the initial sample is selected from a file of cards which record information about stratum membership whereas the subsample entails a personal interview.

6. DOMAIN ESTIMATION

So far we have discussed the estimation of population totals, means, ratios and proportions. However, survey data are often used for analytical purposes as well, i.e., for investigating relationships among variables. Such analyses often involve the comparison of means of certain subgroups (or domains) of the population. The domains are usually well defined but it is not known until after the enumeration as to which of the domains a particular unit belongs so that the sample size within a domain is a random variable. The number of units in a domain (i.e., the domain size) is usually unknown and may well be another parameter of interest. In the example concerning reading habits of freshmen, two domains of particular interest might be male and female students whose first language is other than English.

We give a simple method for the estimation of domain totals and means which is applicable to any sampling design. The technique utilizes only the standard formulae pertinent to the estimation of a population total or ratio. Appropriate to any particular sampling design, we have considered the estimation of the population total $Y = Y(y_j)$ by $\hat{Y} = \hat{Y}(y_j)$ and its variance $V(\hat{Y}) = V(y_j)$ by $v(\hat{Y}) = v(y_j)$. Since these operators (which depend on the sample design) are applicable to any set of characters y_j, the total ${}_iY$ for the ith domain D_i can be estimated by attaching the character ${}_iy_j$ to the ith unit in the population, where ${}_iy_j = y_j$ if the jth unit belongs to D_i, ${}_iy_j = 0$ otherwise. Because ${}_iY = Y({}_iy_j)$, it immediately follows that ${}_i\hat{Y} = \hat{Y}({}_iy_j)$, $V({}_i\hat{Y}) = V({}_iy_j)$, and $v({}_i\hat{Y}) = v({}_iy_j)$. If $\hat{Y}$ and $v(\hat{Y})$ are unbiased for Y and $V(\hat{Y})$ respectively, then the corresponding domain estimators ${}_i\hat{Y}$ and $v({}_i\hat{Y})$ will also be unbiased for ${}_iY$ and $V({}_i\hat{Y})$.

The domain size ${}_iN$ may be expressed in terms of the count variable ${}_ia_j$ where ${}_ia_j = 1$ if the jth unit belongs to D_i, ${}_ia_j = 0$ otherwise. Thus ${}_iN = Y({}_ia_j)$ and hence ${}_i\hat{N} = \hat{Y}({}_ia_j)$, $V({}_i\hat{N}) = V({}_ia_j)$, $v({}_i\hat{N}) = v({}_ia_j)$.

The domain mean ${}_i\bar{Y} = {}_iY/{}_iN$ can be estimated by ${}_i\hat{\bar{Y}} = \hat{Y}({}_iy_j)/\hat{Y}({}_ia_j)$, a ratio estimator. Using the approximate variance formulae for a ratio, we get $V({}_i\hat{\bar{Y}}) \doteq V({}_iy_j - {}_i\bar{Y}{}_ia_j)/N^2$ and $v({}_i\hat{\bar{Y}}) \doteq$

$v({}_iy_j - {}_i\hat{\bar{Y}}_i a_j)/\hat{N}^2$. Note that ${}_iy_j - {}_i\bar{Y}_i a_j$ takes the value $y_j - {}_i\bar{Y}$ if the jth unit is in D_i, and zero otherwise.

The difference between two domain means, say ${}_1\bar{Y} - {}_2\bar{Y}$, is estimated unbiasedly by ${}_1\hat{\bar{Y}} - {}_2\hat{\bar{Y}}$ with variance

$$V({}_1\hat{\bar{Y}} - {}_2\hat{\bar{Y}}) \doteq V[({}_1y_j - {}_1\bar{Y}_1 a_j)/{}_1N - ({}_2y_j - {}_2\bar{Y}_2 a_j)/{}_2N],$$

and variance estimator

$$v({}_1\hat{\bar{Y}} - {}_2\hat{\bar{Y}}) \doteq v[({}_1y_j - {}_1\hat{\bar{Y}}_1 a_j)/{}_1\hat{N} - ({}_2y_j - {}_2\hat{\bar{Y}}_2 a_j)/{}_2\hat{N}].$$

One could set up a confidence interval for ${}_1\bar{Y} - {}_2\bar{Y}$ by assuming that the standardized statistic $[({}_1\hat{\bar{Y}} - {}_2\hat{\bar{Y}}) - ({}_1\bar{Y} - {}_2\bar{Y})]/\sqrt{v({}_1\hat{\bar{Y}} - {}_2\hat{\bar{Y}})}$ is approximately normal with zero mean and unit variance.

As an example, consider a simple random sample of size n. Because $\hat{Y}(y_j) = N\sum_{j\in s} y_j/n$, we obtain ${}_i\hat{Y} = \hat{Y}({}_iy_j) = N\sum_{j\in s} {}_iy_j/n = N({}_iy/n)$, where ${}_iy$ is the sample total for D_i. The domain size ${}_iN$ and domain mean ${}_i\bar{Y}$ are estimated by ${}_i\hat{N} = N({}_in/n)$ and the ratio estimator ${}_i\hat{\bar{Y}} = {}_iy/{}_in = {}_i\bar{y}$, where ${}_in$ is the number of sample units falling into D_i. Moreover, the formulae for $v({}_i\hat{Y})$, $v({}_i\hat{\bar{Y}})$ and $v({}_1\hat{\bar{Y}} - {}_2\hat{\bar{Y}})$ reduce to

$$v({}_i\hat{Y}) = \frac{N(N-n)}{n(n-1)}\left[({}_in - 1){}_is^2 + \left(\frac{1}{{}_in} - \frac{1}{n}\right){}_iy^2\right], \tag{22}$$

$$v({}_i\hat{\bar{Y}}) \doteq \left(\frac{1}{n} - \frac{1}{N}\right)\frac{n^2({}_in - 1)}{{}_in^2(n-1)}\,{}_is^2, \tag{23}$$

$$v({}_1\hat{\bar{Y}} - {}_2\hat{\bar{Y}}) \doteq \left(\frac{1}{n} - \frac{1}{N}\right)\left(\frac{n^2}{n-1}\right)\left[\frac{({}_1n - 1){}_1s^2}{{}_1n^2} + \frac{({}_2n - 1){}_2s^2}{{}_2n^2}\right],$$

where ${}_is^2 = \sum_{j\in D_i}(y_j - {}_i\bar{y})^2/({}_in - 1)$. The first term of (22) corresponds to the case where the domain sizes are known; the second term may be regarded as the contribution to the estimated variance due to the estimation of the domain sizes. Formula (23) (which involves no such additional term) is, approximately, the familiar estimated variance of a mean of a simple random sample of size ${}_in$ from the domain D_i (replacing the unknown ${}_iN$ by its estimator ${}_i\hat{N}$).

The general formulae for variances and estimated variances in the case of more complex designs are often lengthy when spelled out in detail. However, for numerical evaluation we need only apply the standard formulae for the population parameters by setting $y_j = 0$ whenever unit j is not in D_i.

If the domain sizes ${}_iN$ are known, then an improved estimator of the domain total ${}_iY$ is given by ${}_i\tilde{Y} = {}_iN_i\hat{\bar{Y}}$ provided ${}_in > 0$. Note that ${}_i\tilde{Y}$ resembles the estimator of a post-stratum total. Under certain conditions ${}_i\tilde{Y}$ can further be improved by the method of synthetic estimation, particularly when ${}_in$ is small. For example, a simple synthetic estimator of ${}_iY$ is obtained by deflating $\hat{Y}$ by the factor ${}_iN/N$, namely ${}_i\hat{Y}^* = ({}_iN/N)\hat{Y}$, where $\hat{Y}$ is the usual estimator of Y. The bias of ${}_i\hat{Y}^*$ is $B = \{({}_iN/N) - ({}_iY/Y)\}Y$, which is small if ${}_iY/Y \doteq {}_iN/N$. Then ${}_i\hat{Y}^*$ could lead to a substantially smaller MSE than either ${}_i\tilde{Y}$ or ${}_i\hat{Y}$.

7. OPTIMAL DESIGNS

The optimal design of a sample survey involves the determination of sample sizes, sample allocations, etc., which will either maximize the precision of an estimator for a specified survey cost or minimize cost subject to a prescribed precision. This requires some knowledge of population parameters (either from prior experience or from pilot studies), and the form of variance and cost functions. We consider optimal designs in a few simple cases.

7.1. Simple random sampling. Suppose Y is to be estimated by $\hat{Y}$ such that

$$\Pr\left[\left|\frac{\hat{Y} - Y}{Y}\right| \leqslant d\right] = 1 - 2\alpha, \tag{24}$$

where d is the permissible margin of error and 2α is the risk (probability) that the actual error will exceed d. For SRS, noting that $\hat{Y}$ is approximately normally distributed for large n, we get $d = z_\alpha\sqrt{V(\hat{Y})}/|Y| = z_\alpha C(\hat{Y})$, where z_α is the $100(1 - \alpha)$ percentile of the standard normal distribution. Hence the desired sample size satisfying (24) is given by $n = n_0/[1 + (n_0/N)]$, where $n_0 = z_\alpha^2 C_y^2/d^2$ and C_y is the population coefficient of variation. Empirically, C_y has been found to be relatively stable over time and over related characters. It is therefore possible in practice to determine the desired sample size by using a past estimate of C_y or the C-value of some related character. If no such estimate is available, one could determine n approximately by drawing the sample in two steps: (i) a small sample of size n_1 is taken and C_y^2 is estimated by $s_1^2/\bar{y}_1^2$, where $\bar{y}_1$ and s_1^2 are the mean and variance computed from the first step sample; (ii) a sample of size $n - n_1$, where $n = z_\alpha^2 s_1^2/(\bar{y}_1^2 d^2)$, is enumerated at the second step.

In surveys dealing with multiple characters, it is often necessary to employ some compromise value of n since the required value may change appreciably from item to item.

7.2. A class of optimal designs. For many of the survey designs that we have encountered, the variance (or MSE) of the estimator of Y is of the form

$$V = A_0 + \sum_1^k \frac{A_i}{r_i}, \tag{25}$$

where the r_i's are functions only of the sample sizes, k is a positive integer depending upon the sample design, and the A_i's depend upon population parameters but not on the r_i's. The cost (or expected cost) function is assumed to be linear in the r_i's, that is

$$C = c_0 + \sum_1^k c_i r_i, \tag{26}$$

where the c_i's are fixed cost constants. The optimal r_i values are obtained by minimizing either V subject to a prescribed value of C or C for a fixed V. Both of these optimizations are equivalent to the problem of minimizing the product

$$(V - A_0)(C - c_0) = \left(\sum A_i/r_i\right)\left(\sum c_i r_i\right). \tag{27}$$

Applying the familiar Cauchy-Schwarz inequality to the right-hand side of (27) gives

$$(V - A_0)(C - c_0) \geqslant \left[\sum (A_i c_i)^{1/2}\right]^2, \tag{28}$$

with equality if and only if $A_i^{1/2}/r_i^{1/2} \propto (c_i r_i)^{1/2}$, i.e., $r_i = \lambda(A_i/c_i)^{1/2}$, where λ is a constant to be determined from the initial conditions.

If C is fixed, then $C - c_0 = \sum c_i r_i = \lambda \sum (A_i c_i)^{1/2}$, or $\lambda = (C - c_0)/\sum (A_i c_i)^{1/2}$. Hence the optimal value of r_j is given by

$$r_j = (C - c_0)\left(\frac{A_j^{1/2}}{c_j^{1/2}}\right) \Big/ \sum (A_i c_i)^{1/2}, \qquad j = 1, \ldots, k. \tag{29}$$

On the other hand, if V is fixed, then $V - A_0 = \sum A_i/r_i = \sum (A_i c_i)^{1/2}/\lambda$, and the optimal r_j value is

$$r_j = \frac{\sum (A_i c_i)^{1/2}(A_j/c_j)^{1/2}}{V - A_0}.$$

It follows from (28) that the minimum values of V for prescribed C and of C for fixed V are, respectively,

$$V_{\text{Min}} = A_0 + \frac{[\sum (A_i c_i)^{1/2}]^2}{C - c_0},$$

$$C_{\text{Min}} = c_0 + \frac{[\sum (A_i c_i)^{1/2}]^2}{V - A_0}.$$

As an example, consider stratified SRS. The variance of $\hat{Y}_{\text{st}}$ is of the form (25) with $k = L$, $r_i = n_i$, $A_0 = -\sum N_h S_h^2$ and $A_h = N_h^2 S_h^2$, $h = 1, \ldots, L$. Assuming a linear cost function of the form (26), where c_0 is the overhead cost and c_h is the cost per unit in stratum h, the optimal values of n_h for a fixed cost, from (29), are given by

$$n_h = \frac{(C - c_0)N_h S_h/c_h^{1/2}}{\sum N_h S_h c_h^{1/2}}, \qquad h = 1, \ldots, L. \tag{30}$$

In the special case where $c_h = c$ for all h, (30) reduces to the well-known Neyman allocation:

$$n_h = n \frac{N_h S_h}{\sum N_h S_h},$$

where $n = (C - c_0)/c$ is the fixed total sample size.

As another example, in two-phase sampling for ratio estimation, it follows from (20) that $V(\hat{Y}_{\text{rd}})$ is of the form (25) with $k = 2$, $A_0 = -NS_y^2$, $A_1 = N^2(S_y^2 - 2RS_{yx} + R^2 S_x^2)$, $A_2 = N^2(2RS_{yx} - R^2 S_x^2)$,

$r_1 = n$ and $r_2 = n'$. In the cost function (26), c_0 is the overhead cost, c_1 the cost per unit of observing y and c_2 that for x.

In a two-stage design with psu's selected PPS with replacement and secondaries by SRS, $V(\hat{\hat{Y}}_{\text{pps}})$ is given by (19). This has the structure of (25) with $k = M + 1$, $A_0 = 0$, $A_1 = \sum p_i(Y_i/p_i - Y)^2 - \sum M_i S_i^2/p_i$, $A_{i+1} = M_i^2 S_i^2/p_i$, $r_1 = m$ and $r_{i+1} = mm_i$, $i = 1, \ldots, M$. The expected cost C for this design is

$$C = c_0 + c_1 m + c_2 E\left(\sum_1^m m_i\right) = c_0 + c_1 m + c_2 \sum_1^M (mm_i) p_i,$$

where c_0 is the overhead cost, c_1 the cost per psu and c_2 the cost per secondary.

A linear cost function may not be realistic in all survey situations. For example, if the cost of travelling between psu's in the above two-stage design is substantial, then a more realistic expected cost function might be

$$C = c_0 + c^*\sqrt{m} + c_1 m + c_2 \sum_1^M (mm_i) p_i,$$

where $c^*\sqrt{m}$ is the total cost of travelling between the m psu's in the sample. Such cost functions cannot be handled using the Cauchy-Schwarz inequality, but the calculus Lagrange multiplier technique may be employed.

7.3. Sampling for multiple characters. Nearly all large scale surveys involve the study of multiple characters so that the optimal design problem becomes more complicated. Suppose there are p characters with associated variance formulae of the form

$$V_j = A_{0j} + \sum_{i=1}^k A_{ij}/r_i, \qquad j = 1, \ldots, p,$$

where $A_{0j}, A_{1j}, \ldots, A_{kj}$ are constants associated with the jth character which depend upon population parameters. One could then minimize the total cost $c_0 + \sum c_i r_i$ subject to $V_j \leqslant V_{0j}$, $j = 1, \ldots, p$, (and subject to other restrictions on the r_i's as well, if any), where V_{0j} is the prescribed variance for the jth character. Note that inequality signs have been introduced into the variance constraints because the minimum cost solution could yield variances that are actually smaller than the prescribed values for some of the characters. By transforming the r_i to $l_i = 1/r_i$, the minimization problem is reduced to a standard convex programming problem with a separable objective function: minimize the convex function $\sum c_i/l_i$ subject to p linear constraints $A_{0j} + \sum_i A_{ij} l_i \leqslant V_{0j}, j = 1, \ldots, p$.

Other constraints on the l_i's may also be introduced. For instance, in the case of stratified SRS, the additional conditions $n_h \geqslant 2$ (i.e., $l_h \leqslant 1/2$), $h = 1, \ldots, L$, ensure unbiased variance estimation. Huddleston, Claypool and Hocking (1970) have developed a computer program to handle convex programming problems that are encountered in sample surveys.

7.4. Optimization in double sampling for stratification. The variance of $\hat{Y}_d = N \sum n'_h \bar{y}_h / n'$, given by (21), may be written as

$$V(\hat{Y}_d) = A_0 + A_1/n' + \sum_1^L A_{2h}/m_h, \tag{31}$$

where $m_h = n' v_h$, $0 < v_h \leqslant 1$. If c_1 denotes the cost per unit of classifying the first-phase sample into strata and c_2 is the cost per unit of recording the y-values, then the expected total cost of the survey is

$$C = c_0 + c_1 n' + c_2 E\left(\sum_1^L n_h\right) = c_0 + c_1 n' + c_2 \sum_1^L W_h m_h. \tag{32}$$

The optimal values of n' and v_h are obtained by minimizing (31) subject to (32) with the additional constraints

$$m_h \leqslant n', \qquad h = 1, \ldots, L. \tag{33}$$

This is a convex programming problem. However, by exploiting the special structure of the inequality constraints (33), it may be shown that the following simple sequential algorithm leads to the optimal solution: (i) minimize (31) subject to (32), ignoring the constraints (33); this can be done using the Cauchy-Schwarz inequality; if each resulting $v_h = m_h/n' \leqslant 1$, then the solution is already optimal; (ii) if, however, the largest of the $\nu_h > 1$, set it equal to unity and repeat the minimization process with the remaining ν_h's; (iii) repeat until all of the ν_h's obtained are less than or equal to one.

7.5. Optimal stratification. Suppose that x is a concomitant character which is correlated with the character of interest y and that all N values of x are known. The problem is to partition the range of x, $a \leqslant x \leqslant b$, into L strata $[a, x_1), [x_1, x_2), \ldots, [x_{L-1}, b]$ so that the variance of the estimator of Y is minimized, by assuming some relationship between y and x.

To simplify the discussion, assume stratified SRS with stratification based on the values of the y character itself. Let $f(y)$ denote the density function of the y variable with $f(y) = 0$ if y does not belong to $[a, b]$. Ignoring the fpc's, the variance of $\hat{Y}_{st}$ may be written as

$$V(\hat{Y}_{st}) = N^2 \sum_1^L W_h^2(y_{h-1}, y_h)\sigma_h^2(y_{h-1}, y_h)/n_h,$$

where

$$W_h(y_{h-1}, y_h) = \int_{y_{h-1}}^{y_h} f(y)\, dy$$

is the proportion of the population units that fall in the stratum $[y_{h-1}, y_h)$, and

$$\bar{Y}_h = \frac{1}{W_h(y_{h-1}, y_h)} \int_{y_{h-1}}^{y_h} yf(y)\, dy,$$

$$\sigma_h^2 = \frac{1}{W_h(y_{h-1}, y_h)} \int_{y_{h-1}}^{y_h} (y - \bar{Y}_h)^2 f(y)\, dy.$$

One could assume either proportional or optimum allocation for the n_h's.

The problem of optimal stratification outlined above has the following general structure: find the set $(y_1, \ldots, y_{L-1})$ which minimizes the objective function $\sum_1^L f_j(y_{j-1}, y_j)$ subject to $y_0 = a \leqslant y_1 \leqslant y_2 \leqslant \cdots \leqslant y_{L-1} \leqslant y_L = b$. Bühler and Deutler (1975) obtained a global optimum solution using a dynamic programming algorithm. One advantage of their solution is that no differentiability assumptions on the objective function are required (unlike some earlier solutions).

A simple approximate solution may be obtained by choosing the stratification points to yield equal intervals on the cumulatives of $\sqrt{f(y)}$ (see Dalenius and Hodges (1959)).

8. NONSAMPLING ERRORS

In the theory developed so far it has been assumed that the true y-values, viz., the y_j^*'s, associated with the sampled units can be obtained exactly. In practice, however, it may not be possible to observe the y^*-values for some of the selected units due to nonresponse or the observed y-values may be subject to response or measurement errors (i.e., $y_j \neq y_j^*$). Both types of nonsampling errors could well occur in the same survey. We first discuss some methods for handling nonresponse.

8.1. The nonresponse problem. Hansen and Hurwitz (1946) describe a technique for dealing with the nonresponse that occurs when n' individuals selected by simple random sampling from a population of size N are sent mail questionnaires. The population may be conceived of as consisting of those individuals who, if contacted by mail, would respond (the response stratum) and those who would fail to respond (the nonresponse stratum). A subsample of size $n_2 = n_2'/k$ is selected by SRS from the $n_2' = n' - n_1'$ nonrespondents and subjected to more costly personal interviews. This procedure clearly is a special case of double sampling for stratification (Subsection 5.6) with $L = 2$, $\nu_1 = 1$ and $\nu_2 = 1/k$. An unbiased estimator of the population total, Y, therefore is given by

$$\hat{Y} = \frac{N}{n'}(n_1'\bar{y}_1 + n_2'\bar{y}_2),$$

with variance

$$V(\hat{Y}) = N^2\left[\left(\frac{1}{n'} - \frac{1}{N}\right)S^2 + \frac{(k-1)}{n'}W_2S_2^2\right],$$

where $\bar{y}_1$ and $\bar{y}_2$ are, respectively, the means of the responses from the mail and the personal interview surveys, S^2 is the population mean square, S_2^2 is the mean square in the nonresponse stratum, and W_2 is the weight of this stratum.

If c' is the unit cost of the first attempt, c_1 the unit cost of processing the data from the response stratum and c_2 the unit cost of interviewing a nonrespondent and processing the questionnaire, then the expected survey cost is

$$C^* = n'(c' + W_1c_1) + (n'/k)W_2c_2,$$

where $W_1 = 1 - W_2$ is the weight of the response stratum. Minimization of C^* subject to a prescribed value V_0 of $V(\hat{Y})$ yields

$$k_{\text{opt}} = \left[\frac{c_2(S^2 - W_2S_2^2)}{S_2^2(c' + W_1c_1)}\right]^{1/2},$$

$$n'_{\text{opt}} = \frac{N^2[S^2 + (k_{\text{opt}} - 1)W_2S_2^2]}{V_0 + NS^2}.$$

If $k_{\text{opt}} < 1$, set $k_{\text{opt}} = 1$.

As an illustration, suppose $n' = 1000$, $W_2 = 0.4$, $c' = \$0.30$, $c_1 = \$1.20$, $c_2 = \$13.50$, $S^2 = S_2^2$ and that the precision desired is that which would be given by a simple random sample of 1,000 units with a 100% response. Then $k_{\text{opt}} = 2.818$ and $n'_{\text{opt}} = 1727$, assuming that N is large. Thus 1727 questionnaires should be mailed out and 35.5% of the nonrespondents interviewed personally. The expected total cost is \$5,071.

Politz and Simmons (1949, 1950), following a suggestion by Hartley (1946), developed an ingenious method for eliminating callbacks in a personal interview survey. If p_i is the probability that the ith person is at home at the time of the enumerator's call, given that he is selected in a simple random sample of size n (the visit being made at a random time within interviewing hours), then the overall probability of an interview with person i is $\pi_i = np_i/N$ and the Horvitz-Thompson estimator of Y is $(N/n) \sum y_i/p_i$, the sum being taken over the responding units. The p_i's are estimated by instructing enumerators to ask respondents whether or not they were at home on, say, the five preceding interviewing days. The bias ratio of the resulting estimator is not likely to be large. Interviews are often held in the evening hours when most people are at home.

In many sample surveys time and budget limitations do not permit a follow-up of the nonrespondents. Even when callbacks are feasible, there is usually a hard core who either refuse to be interviewed under any circumstances or who are not available at any time. One approach that is sometimes used in an attempt to reduce the bias of nonresponse is to obtain substitutes from the population with characteristics similar to those of the nonrespondents (perhaps from the same stratum or psu). The method has the advantage of maintaining the original sample size, but the total sample (original respondents plus substitutes) does not constitute a probability sample from the entire population since the substitutes are obtained from the response stratum. Alternately, the sample data can be adjusted at the analysis stage to compensate for nonresponse. The simplest approach is to ignore the nonrespondents and deal with the respondents as if they constituted a probability sample of reduced size from the entire population. This technique essentially imputes the mean of the respondents to the nonrespondents.

Even though an individual is classified as a respondent, the questionnaire itself may have one or more fields that are incomplete. Item nonresponse also occurs at the editing stage when a questionnaire entry is declared to be inconsistent with other entries or to be erroneous and is therefore rejected. Sometimes an approximate value can be imputed for a missing response on the basis of information provided elsewhere on the same questionnaire. So-called "cold deck" and "hot deck" imputation procedures are also popular at the present time. The former uses observations derived from the cells of a one-way (or multi-way) classification of the values obtained from a previous survey of a similar population. The cold deck distribution is prepared in advance for each cell and stored in the computer memory. When a missing entry is detected for a particular respondent to the current survey, the values of the appropriate completed items are used to identify a cell of the table. An imputed value is then established by selecting either randomly or systematically a value from the cold deck responses belonging to the relevant cell. A hot deck imputation, while more costly, utilizes a cross-classification of the current responses. The initial values are supplied by a cold deck but responses from the new (hot) deck are substituted as the current file is processed. Because there is often some special ordering to the respondents, the value imputed is usually that of the last respondent recorded in the corresponding cell.

8.2. Response errors. In Section 3 we presented a general survey error model in which the survey process was regarded as conceptually repeatable. The true population total Y^* was estimated by $\hat{Y}_t$, calculated from the observed y_{it}-values on trial t. Letting y_i be the expected value of y_{it} over all trials and samples containing unit i and $\hat{Y}$ the value of $\hat{Y}_t$ with y_{it} replaced by y_i, the MSE of $\hat{Y}_t$ (over all trials and samples) was expressed as the sum of four components:

$$\begin{aligned}\text{MSE}(\hat{Y}_t) &= E(\hat{Y}_t - \hat{Y})^2 + E(\hat{Y} - Y)^2 \\ &\quad + 2E[(\hat{Y}_t - \hat{Y})(\hat{Y} - Y)] + (Y - Y^*)^2, \qquad (34)\end{aligned}$$

where $Y = \sum_{i=1}^{N} y_i$. The first term on the right-hand side of (34) is the response variance, the second term the sampling variance, the third term an interaction component and the last term the squared response bias.

To simplify the ensuing discussion, we assume SRS and a negligible response bias, i.e., $Y = Y^*$. The response variance may be written as

$$E(\hat{Y}_t - \hat{Y})^2 = \frac{N^2}{n}[1 + (n-1)\rho]\sigma_R^2.$$

Here $\sigma_R^2 = E(y_{jt} - y_j)^2$ is called the simple response variance and $\rho\sigma_R^2 = E(y_{jt} - y_j)(y_{kt} - y_k)$, $j \neq k$, is the correlated response variance representing the variability induced by any correlations that might exist among the response deviations for different units for a given survey trial, where E denotes the expectation operator taken over all trials and samples.

For large n, the correlated response contribution, namely $N^2(n-1)\rho\sigma_R^2/n$, could dominate the simple response variance term $N^2\sigma_R^2/n$ as well as the sampling error contribution, $N^2\{(1/n) - (1/N)\}S_y^2$. In the case of a census in which the sampling error is zero, the correlated response component could be so large that the census total MSE might exceed that of a well-planned survey which used intensive procedures to reduce nonsampling errors. In interview surveys, much of the correlated response variance arises from enumerators who, through their own attitudes, interpretation of questions, or interviewing techniques, tend to induce a positive correlation among the response deviations in their assignment. Recognizing this, censuses in such countries as the United States, Canada and Sweden now feature self-enumeration procedures for which ρ usually is negligible.

In order to estimate the simple response variance, it is necessary to have replicated measurements on the sample units, each of the measurements being taken under similar circumstances. In the case of two such measurements, an approximately unbiased estimator of σ_R^2 is given by the statistic $g = \sum (y_{j1} - y_{j2})^2/(2n)$. In the case of different agents in the two trials (e.g., different data coders), another statistic is also constructed, namely $C = (\bar{y}_1 - \bar{y}_2)^2/2$ with expectation $E(C) = \sigma_R^2(1 + (n - 1)\rho)/n$. Estimates of both σ_R^2 and $\rho\sigma_R^2$ can then be calculated using the observed g and C values.

The method of interpenetrating subsamples due to Mahalanobis (1946) may be employed in the situation where it is either not possible or not practical to obtain repeated observations on the same unit. Assume that the sample of size n is divided into k subsamples each containing $\bar{n} = n/k$ units. If k interviewers are assigned at random to the k subsamples (one per subsample) and if there is no correlation between the measurement errors of different enumerators, then the variance of the estimator $\hat{Y}_I = N \sum_{i=1}^{k} \bar{y}_{it}/k$ is

$$V(\hat{Y}_I) = \frac{N^2}{n}\{S_y^2 + \sigma_R^2[1 + (\bar{n} - 1)\rho]\}, \tag{35}$$

where $\bar{y}_{it}$ is the mean for the ith subsample and ρ is the correlation among the response deviations obtained by the same enumerator. An unbiased estimator of $V(\hat{Y}_I)$ is obtained by setting up a sample analysis of variance. The total sum of squares is decomposed into a sum of squares between interviewers within subsamples, denoted by $(k - 1)s_b^2$ with $k - 1$ degrees of freedom and expectation $(k - 1)(S_y^2 + \sigma_R^2[1 + (\bar{n} - 1)\rho])$, and a sum of squares within interviewers, denoted by $k(\bar{n} - 1)s_w^2$ with $k(\bar{n} - 1)$ degrees of freedom and expectation $k(\bar{n} - 1)[S_y^2 + \sigma_R^2(1 - \rho)]$. It follows from

(35) that $N^2 s_b^2/n$ is an unbiased estimator of $V(\hat{Y}_I)$. Similarly, the correlated response variance $\rho\sigma_R^2$ is unbiasedly estimated by $(s_b^2 - s_w^2)/\bar{n}$. No estimate of the simple response variance σ_R^2 can, however, be obtained.

A number of extensions and refinements of the above designs are available in the literature. An excellent discussion is given by Bailar and Dalenius (1970).

Since the response bias involves the true population total Y^*, it cannot be unbiasedly estimated from the observed values in a census or survey. Some countries employ a small-scale post-enumeration survey after a census in an attempt to estimate the response bias. A sample of units is subjected to intensive interviews by highly qualified enumerators using detailed survey procedures to obtain (hopefully) the true y^*-values, an effort that would be far too costly in a census or large-scale survey. By matching the sample results with census records, estimates of the response bias can be derived.

9. SOME RECENT DEVELOPMENTS

The ever increasing reliance on analytical methods in the social, educational and health sciences, among others, has focused attention on a variety of problems in the design, analysis and practice of sample surveys. In order to impart some idea of the types of problems receiving attention and the research being conducted, we present some developments in three important areas: (i) the analysis of complex survey data; (ii) the use of two or more frames; (iii) methods for handling sensitive questions.

9.1. Analyzing complex survey data. Many surveys now incorporate complex strategies involving such features as stratification and multistage sampling coupled with detailed statistical analyses, e.g., regression and correlation analysis, analysis of variance and chi-square contingency tests of association. Due to the lack of satisfactory methods for handling such complex survey data, practitioners have often been forced to rely upon standard statistical methods based on the assumption of simple random sampling leading, possibly, to serious errors in inference. Recently, some ingenious methods have been proposed that take the complexities into account. Two such techniques are named "balanced repeated replication" (BRR) and the "jackknife."

The BRR method, proposed by McCarthy (1966), essentially involves the construction of a set of estimates $\hat{\theta}_1, \ldots, \hat{\theta}_k$ of a population parameter θ from a selected sample such that the standard error of the full-sample estimate $\hat{\theta}$ is simply obtained from $v_B(\hat{\theta}) = (1/k)\sum_1^k (\hat{\theta}_i - \hat{\theta})^2$. This set of estimates is so chosen as to make $v_B(\hat{\theta})$ stable. To illustrate the technique, consider the case of a stratified design with two units per stratum selected by SRSWR, the parameter of interest being $\theta = Y$. If y_{h1} and y_{h2} are the sample observations from stratum h, the customary estimator of $V(\hat{Y}_{st})$ is given by $v(\hat{Y}_{st}) = \frac{1}{4}\sum N_h^2 d_h^2$, where $d_h = y_{h1} - y_{h2}$. In this case the BRR method forms a set of half-samples by selecting one unit from each stratum in such a way that $v_B(\hat{Y}_{st}) \equiv v(\hat{Y}_{st})$. The ith half-sample estimate, $\hat{Y}_{st,i}$, is formed by letting $\delta_{h1}^{(i)} = 1$ if unit $h1$ is in the ith half-sample and zero otherwise, and setting $\delta_{h2}^{(i)} = 1 - \delta_{h1}^{(i)}$:

$$\hat{Y}_{st,i} = \sum N_h(\delta_{h1}^{(i)} y_{h1} + \delta_{h2}^{(i)} y_{h2}).$$

If one selects all $k = 2^L$ possible half-samples, it may then be verified that

$$v_B(\hat{Y}_{st}) = \sum_i (\hat{Y}_{st,i} - \hat{Y}_{st})^2/2^L \equiv v(\hat{Y}_{st}).$$

But the full set of 2^L half-samples could be very large. The BRR method selects a balanced subset of $m < 2^L$ half-samples such that

$$v_B(\hat{Y}_{st}) = \sum_{i=1}^{m} (\hat{Y}_{st,i} - \hat{Y}_{st})^2/m \equiv v(\hat{Y}_{st}).$$

This is accomplished by noting that $v_B(\hat{Y}_{st})$ may be written, with $\delta_h^{(i)} = \delta_{h1}^{(i)} - \delta_{h2}^{(i)}$, as

$$v_B(\hat{Y}_{st}) = \frac{1}{4}\sum N_h^2 d_h^2 + \frac{1}{2m}\sum_{h<h'}\sum \left(\sum_i^m (\delta_h^{(i)}\delta_{h'}^{(i)}\right) N_h N_h' d_h d_h',$$

and hence by choosing a set of half-samples such that $\sum_i^m \delta_h^{(i)}\delta_{h'}^{(i)} = 0$ for all $h < h'$. Such half-samples are readily obtained from Hadamard

	Stratum 1	2	3	4	5	6	7	8
Half-sample 1	+1	−1	−1	+1	−1	+1	+1	−1
2	+1	+1	−1	−1	+1	−1	+1	−1
3	+1	+1	+1	−1	−1	+1	−1	−1
4	−1	+1	+1	+1	−1	−1	+1	−1
5	+1	−1	+1	+1	+1	−1	−1	−1
6	−1	+1	−1	+1	+1	+1	−1	−1
7	−1	−1	+1	−1	+1	+1	+1	−1
8	−1	−1	−1	−1	−1	−1	−1	−1

FIG. 1. Hadamard matrix of order eight.

matrices $\mathbf{H}_m$ of order m with elements $+1$ or -1 such that $\mathbf{H}_m^T\mathbf{H}_m = m\mathbf{I}_m$ where a "$+1$" denotes the first sampled unit, a "-1" the second sampled unit from a stratum and $\mathbf{I}_m$ is the identity matrix of order m. Plackett and Burman (1946) have given a method for constructing such matrices when m is a multiple of 4. Figure 1 displays a Hadamard matrix for $m = 8$ where the rows identify the half-samples and any set of 4, 5, 6 or 7 columns (selected from the first seven columns) for the 4, 5, 6 or 7 strata cases define a set of $k = 8$ half-samples with the property $\sum_i^m \delta_h^{(i)}\delta_{h'}^{(i)} = 0$ for all $h < h'$.

As an application of BRR, consider the problem of testing the independence of two criteria of classification in an $r \times c$ contingency table from a stratified cluster sample. Let $P_{ijh} = N_{ijh}/N$ be the proportion of population units in stratum h falling in cell (i, j) of the table. Denoting the marginal totals of the classification by $P_{ij.}$, $P_{i..}$, and $P_{.j.}$, the hypothesis may be expressed as

$$H_0 : P_{ij.} = (P_{i..})(P_{.j.}), \qquad i = 1, \ldots, r; j = 1, \ldots, c.$$

Note that the stratification employed is of no intrinsic interest as an additional variable when testing the independence between the two classification variables. Suppose that two psu's are selected from each stratum by SRS and that subsampling is performed to yield two independent estimates $\hat{P}_{ijh1}$ and $\hat{P}_{ijh2}$ of P_{ijh}. If a set of m

balanced half-samples is selected, then an estimate of $P_{ij.}$ from the uth half-sample, is

$$\hat{P}_{ij.u} = \sum_h (\delta_{h1}^{(u)} \hat{P}_{ijh1} + \delta_{h2}^{(u)} \hat{P}_{ijh2}), \qquad u = 1, \ldots, m.$$

From the complement of the uth half-sample we obtain an independent estimate of $P_{ij.}$, namely

$$\tilde{P}_{iju} = \sum_h (\delta_{h2}^{(u)} \hat{P}_{ijh1} + \delta_{h1}^{(u)} \hat{P}_{ijh2}), \qquad u = 1, \ldots, m.$$

Note that under the null hypothesis the statistics

$$\hat{Q}_{ij,u} = \hat{P}_{ij.u} + \tilde{P}_{ij.u} - \hat{P}_{i..u}\tilde{P}_{.j.u} - \tilde{P}_{i..u}\hat{P}_{.j.u},$$
$$i = 1, \ldots, r-1; \quad j = 1, \ldots, c-1,$$

have zero expectation. Hence we can construct a chi-square type statistic

$$X_Q^2 = \bar{\hat{\mathbf{Q}}}^T \hat{\mathbf{D}}_Q^{-1} \bar{\hat{\mathbf{Q}}},$$

where $\bar{\hat{\mathbf{Q}}}$ is the $(r-1)(c-1) \times 1$ column vector with elements

$$\bar{\hat{Q}}_{ij} = \frac{1}{m} \sum_{u=1}^{m} \hat{Q}_{iju},$$

and $\hat{\mathbf{D}}_Q$ is the estimated covariance matrix of the $\bar{\hat{Q}}_{ij}$'s; a formula for $\hat{\mathbf{D}}_Q$ is given by Nathan (1973). The null hypothesis of independence can be tested by comparing the observed value of the X_Q^2 statistic with an upper percentile of the chi-square distribution with $(r-1)(c-1)$ degrees of freedom.

Recently, some extensions to the case $n_h = \bar{n} > 2$ have been reported but much remains to be done on constructing BRR's (both for equal and unequal sample sizes within strata) with a minimum number of replicates. The theory used in the construction of experimental designs will most certainly provide useful tools for developing optimal BRR's.

As in the case of BRR, the jackknife method also involves a set of estimates $\hat{\theta}_1, \ldots, \hat{\theta}_k$ constructed from a chosen sample such that the variance of the full estimate $\hat{\theta}$ is simply obtained from the squared deviations $(\hat{\theta}_i - \hat{\theta})^2$. We illustrate the method for the stratified cluster sample design involving two sample clusters per stratum selected with replacement. Suppose $\hat{\theta}_{hi}$ denotes the estimate of θ obtained from the whole sample after deleting the ith selected cluster from stratum h, $i = 1, 2$. In all, $k = 2L$ such estimates can be constructed. A jackknife variance estimator is then given by

$$v_J(\hat{\theta}) = \frac{1}{2}\sum_h \sum_i (\hat{\theta}_{hi} - \hat{\theta})^2.$$

In the linear case, i.e., $\hat{\theta} = \hat{Y}_{st}$, the variance estimator $v_J(\hat{\theta})$ reduces to $v(\hat{Y}_{st})$, as in the case of BRR.

It is possible to construct a variety of jackknife variance estimators, e.g.,

$$v_{J1}(\hat{\theta}) = \sum (\hat{\theta}_{h1} - \hat{\theta})^2,$$
$$v_{J2}(\hat{\theta}) = \sum (\hat{\theta}_{h2} - \hat{\theta})^2,$$
$$v_{J3}(\hat{\theta}) = \frac{1}{4}\sum (\hat{\theta}_{h1} - \hat{\theta}_{h2})^2,$$

all of which reduce to $v_J(\hat{\theta})$ in the linear case. Similar variance estimators may also be constructed in the BRR situation.

9.2. Multiple frame surveys. One often finds that a sampling frame known to cover all units in the study population is also one in which sampling is costly, whereas other frames (perhaps special lists of units) are less costly to employ but cover only a possibly unknown fraction of the population. In other situations no single frame may encompass the population of interest, but the union of two or more overlapping frames covers all units. Hartley (1974) has developed a general theory for utilizing two frames, A and B, without requiring any prior information about the extent of their overlap.

The units in the two frames could well be of different types. For example, frame A might be a list of hospitals whereas frame B might consist of households with household members as units. The y-value, y_A, of a unit in frame A can be written as $y_A = y_a + y_{ab}$ where y_a and y_{ab} are the contributions from units not covered in frame B and from units included in frame B, respectively. Likewise, $y_B = y_b + y_{ba}$, with a similar interpretation for y_b and y_{ba}. In the above example, y_A might denote the total number of discharges reported by a hospital in a specified reference period. This may be split into y_{ab} = discharges of persons living in households and y_a = discharges of persons not belonging to households but, say, to institutions. Similarly, y_B is the number of discharges from *any* hospital of a member of a household during the reference period, and y_{ba} is the number of discharges from a hospital in the list frame A.

In the special case where frames A and B embrace overlapping fractions of the same population of units, frame A would be split into domains a (units accessible via frame A only) and ab (units accessible via frames A or B) with $y_{ab} = 0$ in domain a and $y_a = 0$ in domain ab. Similarly, in frame B, $y_{ba} = 0$ in domain b (units accessible through frame B only) and $y_b = 0$ in domain ab. Let $Y_A = Y_A(y_A)$ and $Y_B = Y_B(y_B)$ denote the population totals for the two frames and $\hat{Y}_A(y_A, \alpha)$, $\tilde{Y}_B(y_B, \beta)$ their corresponding single frame estimators which are linear functions of the y-values, where α and β designate the design parameters (e.g., sample sizes) associated with frames A and B. The total Y of the study population can be written as

$$Y = Y_a + Y_{ab} + Y_b,$$

where $Y_a = Y_A(y_a)$, $Y_b = Y_B(y_b)$ and $Y_{ab} = Y_{ba} = Y_A(y_{ab}) = Y_B(y_{ba})$. Noting that $\hat{Y}_A(y_{ab}, \alpha)$ and $\tilde{Y}_B(y_{ba}, \beta)$ are both estimating Y_{ab}, a multiple frame estimator of Y is then given by

$$\hat{Y}_m = \hat{Y}_A(y_a, \alpha) + \tilde{Y}_B(y_b, \beta) + p\hat{Y}_A(y_{ab}, \alpha) + q\tilde{Y}_B(y_{ba}, \beta),$$

where $q = 1 - p$, and p is a parameter to be optimized along with α and β. The determination of the optimum values of p, α and β is discussed by Hartley.

Consider the precision of a two frame estimator as compared to that of a single frame estimator under SRS in the special case where frame B is complete and both frames have the same type of unit. For a typical situation with a population variance ratio $\sigma_A^2/\sigma_b^2 = 16$, cost ratio per unit $C_A/C_B = 0.2$, and $N_A/N = 0.8$, the relative efficiency of the single frame B estimator to the multiple frame estimator is 26%. Such substantial gains in efficiency are realized for a wide range of values of the above three ratios.

Besides the considerable increase in precision for a given cost that is likely to be realized with a multiple frame approach as compared to the single frame survey, other advantages of the former are: (i) if the 100% frame B is an expensive frame to sample but other cheaper frames which are all incomplete are available, then a survey in frame B alone may be too costly for the same precision resulting in, nevertheless, often unsuccessful attempts to undertake a single frame survey by "patching" one of the other frames; (ii) if frame B involves an interview survey and A a mail survey, then the subsampling and interviewing of the nonrespondents to the mail questionnaire may be operationally combined with the direct interview operation in B.

9.3. Sampling population graphs. As contrasted to a statistician who treats a population as a set of units with associated characteristics, a sociologist or a biologist often regards it as a system characterized by important interactions and structural positions of its units. Stephan (1969) emphasized the importance of utilizing to a greater extent in the sample design what is known (or knowable) about the population structure.

The term "nexus sampling" refers to sampling procedures that are affected by the network of relationships existing in the population. Graph theory has proven to be a valuable tool for mathematical modelling to take account of the nexus properties of the population. A simple way to introduce structure into a finite population of units is to supplement the variate values by a set of relationships between every pair of units. Thus, for every pair (i, j), either i is related to j, denoted by R_{ij}, or not related, say, $\bar{R}_{ij}$. The relationships may be viewed as a directed graph with units as nodes and the relationships as arcs. If $R_{ij} = R_{ji}$, then the relationships may be treated as an undirected graph. A structure matrix $\mathbf{X}$ with elements $x_{ij} = 1$ if R_{ij} holds, $x_{ij} = 0$ otherwise, permits the use of algebraic techniques to analyze the relationships.

As an illustration, in a class of students $x_{ij} = 1$ might imply that student j regards student i as the class leader and $x_{ij} = 0$ that the student does not. The nodes of the graph are the students and the arcs are the pairs (i, j) where i is the student most respected in the class by j. A sociologist might be interested in investigating the nature of the influence chain on the basis of a sample of students. A value matrix could be defined by giving some numerical measure on, say, a zero to ten scale to indicate the degree to which j respects i. In making inferences about valued graphs, the prior information available about the population graph, the sampling procedure to be used and the working procedures for obtaining data from the sample must be specified.

Goodman (1961) used a sequential sampling procedure termed "snowball sampling" to estimate the number of mutual arcs in a directed population graph known to have exactly one arc from every node. A k-stage snowball sample consists of an initial sample of nodes, a first stage composed of nodes not in the sample but adjacent to some node in the initial sample, a second stage containing nodes not in the initial or first stage samples but adjacent to a node in the first stage sample, and so on to k stages.

The wide availability of computers and the sophistication of computer software have given great impetus to research in the area of statistical inference for population graphs but much remains to be done.

9.4. Randomized response. Warner (1965) developed a method for increasing the cooperation of individuals in a survey involving questions of a sensitive or personal nature. His randomized response technique is based upon the premise that the incidence of non-response or incorrect answers will substantially diminish if the interviewee is permitted to maintain his privacy (which is indeed desirable in its own right). Thus, suppose that every person in a population belongs to one of two mutually exclusive groups, A and B, and that an estimate of the proportion π belonging to group A is required. A randomizing device, such as a spinner, is given to the respondent in order to permit him, unobserved by the enumerator, to select one of the following two questions with probabilities p and $1 - p$ respectively:

Question 1. Do you belong to group A?
Question 2. Do you belong to group B?

The respondent answers either "Yes" or "No" to the chosen question but does not identify which of the two questions was actually picked. Presumably the respondent will answer the selected question truthfully since he cannot be classified with certainty as to his actual group membership (assuming $0 < p < 1$), thereby removing any embarrassment or stigmatization that might otherwise be induced.

Suppose that a simple random sample of n individuals selected from a large population is subjected to the above interviewing procedure and that r reply "Yes" and $n - r$ "No." The probability of the respondent replying "Yes" is $\lambda = p\pi + (1 - p)(1 - \pi)$. The maximum likelihood estimator of λ is $\hat{\lambda} = r/n$, and consequently the maximum likelihood estimator of π (which is also unbiased provided all respondents answer truthfully) is

$$\hat{\pi} = \frac{p - 1}{2p - 1} + \frac{r}{n(2p - 1)},$$

with variance

$$V(\hat{\pi}) = \frac{\pi(1 - \pi)}{n} + \frac{1}{4n}[(2p - 1)^{-2} - 1],$$

provided $p \neq 1/2$. The first component of $V(\hat{\pi})$ is the binomial variance under direct questioning with truthful reporting; the second term is the additional contribution due to the randomizing device. Evidently $V(\hat{\pi})$ decreases as p deviates from $1/2$, but too large or too small a choice of p could lead to response problems since the interviewee might then suspect a violation of the confidentiality supposedly afforded by the device.

The randomized response method may be compared with direct questioning and a 100% response. Let T_A denote the constant probability that any member of group A gives a truthful response and similarly for T_B. Suppose that $\tilde{\pi}$ denotes the proportion of the n sample members that reply "Yes" to the direct question "Do you belong to Group A?". Then $\tilde{\pi}$ is in general biased for π with expectation $E(\tilde{\pi}) = \pi T_A + (1 - \pi)(1 - T_B) = \lambda_D$, and with variance $V(\tilde{\pi}) = \lambda_D(1 - \lambda_D)/n$. Let $E = V(\hat{\pi})/\text{MSE}(\tilde{\pi})$ measure the relative efficiency of the competing approaches. Suppose $T_A = 1.0$, $T_B = 0.9$ and $\pi = 0.6$, representing, perhaps, a voting situation where 10% of the minority population B would fail to divulge their status and would reply "A" rather than "B" to a direct question. With a simple random sample of $n = 1{,}000$, the values of E for $p = 0.6$, 0.7 and 0.8, respectively, are 3.4, 0.85 and 0.37. If $\pi = 0.05$ for a sensitive question A such as "Have you ever suffered from a venereal disease?" with $n = 1{,}000$, $T_A = 0.6$, $T_B = 1.0$, then E assumes the values 3.2, 1.1 and 0.44 for $p = 0.7$, 0.8 and 0.9, respectively. However, the Warner method is not usually suitable for those situations where π is small unless there is substantial deception.

Simmons (see Greenberg *et al.* (1969)) suggested that more truthful responses might be elicited in a potentially stigmatizing or embarrassing situation if one of the two questions asked is nonsensitive, innocuous and unrelated to the sensitive question. For example, the questions might be:

Question 1. Are you currently having an affair outside of your marriage?

Question 2. Were you born in the months of April, May, June or July?

If Question 2 refers to membership in Group Y with proportion π_Y in the population and the sensitive Question 1 to membership in group A with proportion π_A, then the probability of a "Yes" reply is $\lambda = p\pi_A + (1 - p)\pi_Y$, where p is the probability that the sensitive question is selected. If π_Y is known (as it is in this example), then the unbiased maximum likelihood estimator of π_A is

$$\hat{\pi}_A = [\hat{\lambda} - (1 - p)\pi_Y]/p,$$

where $\hat{\lambda}$ is the proportion of "Yes" answers in the sample. The variance of $\hat{\pi}_A$ is $V(\hat{\pi}_A) = \lambda(1 - \lambda)/(np^2)$. If π_Y is unknown, then two samples must be observed; in one sample only the nonsensitive question is presented and in the other sample the unrelated question approach is used.

A host of extensions and refinements have appeared since publication of Warner's initial paper. A series of articles in the *International Statistical Review* (1976) discuss many variants of the procedure.

Much of the previous literature has concentrated on the statistical properties of the randomized response approach with little attention being paid to the degree of privacy provided to the individual by alternative schemes. Lanke (1976), among others, recognized that the larger the conditional probability of an individual belonging to Group A, given a certain response, the greater is the embarrassment caused by giving that answer. He therefore proposed the conditional probability Pr(respondent belongs to group A | respondent has replied "Yes"), denoted by $\Pr(A|\text{yes})$, as a measure of the maximal embarrassment connected with giving a truthful reply. Then one randomized interview method is more protective than another if $\max[\Pr(A|\text{yes}), \Pr(A|\text{no})]$ is smaller for the former method than for the latter. Efficiency comparisons under the restriction of equal jeopardy are then made for a number of approaches.

An alternative to the randomized response method using incomplete block designs has been given by Raghavarao and Federer (1973). An interviewee is asked to provide only the overall total of the individual answers to a mixture of sensitive and neutral questions. From these, estimated means for each of the questions may then be derived.

SELECTED BIBLIOGRAPHY OF TEXTBOOKS

Elementary

1. V. Barnett, *Elements of Sampling Theory*, English Universities Press, London, 1974.
2. W. Mendenhall, L. Ott, and R. L. Scheaffer, *Elementary Survey Sampling*, Wadsworth Publ. Co., Belmont, Calif., 1971.
3. M. J. Slonim, *Sampling in a Nutshell*, Simon and Schuster, New York, 1960.
4. A. Stuart, *Basic Ideas of Scientific Sampling*, Charles Griffin and Co., London, 1962.
5. T. Yamane, *Elementary Sampling Theory*, Prentice-Hall, Englewood Cliffs, N.J., 1967.

Applied

6. W. E. Deming, *Sample Design in Business Research*, Wiley, New York, 1960.
7. M. H. Hansen, W. N. Hurwitz, and W. G. Madow, *Sample Survey Methods and Theory*, vol. I, Wiley, New York, 1953.
8. L. Kish, *Survey Sampling*, Wiley, New York, 1967.
9. D. Raj, *The Design of Sample Surveys*, McGraw-Hill, New York, 1972.
10. M. R. Sampford, *An Introduction to Sampling Theory*, Oliver and Boyd, Edinburgh and London, 1962.
11. F. S. Stephan and P. J. McCarthy, *Sampling Opinions*, Wiley, New York, 1958.
12. F. Yates, *Sampling Methods for Censuses and Surveys*, 3rd ed., Charles Griffin, London, 1960.

Theoretical

13. W. G. Cochran, *Sampling Techniques*, 3rd ed., Wiley, New York, 1977.
14. W. E. Deming, *Some Theory of Sampling*, Wiley, New York, 1950.
15. M. H. Hansen, W. N. Hurwitz, and W. G. Madow, *Sample Survey Methods and Theory*, vol. II, Wiley, New York, 1953.
16. M. N. Murthy, *Sampling Theory and Methods*, Statistical Publ. Co., Calcutta, 1967.
17. H. S. Konijn, *Statistical Theory of Sample Survey Design and Analysis*, North-Holland, London, 1973.
18. D. Raj, *Sampling Theory*, McGraw-Hill, New York, 1968.
19. P. V. Sukhatme and B. V. Sukhatme, *Sampling Theory of Surveys with Applications*, FAO of the United Nations, Rome, 1970.

BIBLIOGRAPHY

1. B. A. Bailar and T. Dalenius, "Estimating the response variance components of the U.S. Bureau of the Census' survey model," *Sankhyā*, **B31** (1969) 341–360.
2. D. L. Bayless and J. N. K. Rao, "An empirical study of stabilities of estimators and variance estimators in unequal probability sampling ($n = 3$ or 4)," *J. Amer. Statist. Assoc.*, **65** (1970) 1645–1667.
3. D. L. Bayless, B. V. Shah, and A. L. Finkner, "Computing sampling errors for two educational assessment surveys," *Bull. of I.S.I.*, XLV, **3** (1973) 47–65.
4. K. R. W. Brewer, "A model of systematic sampling with unequal probabilities," *Australian J. Statist.*, **5** (1963) 5–13.
5. W. Bühler and T. Deutler, "Optimal stratification and grouping by dynamic programming," *Metrika*, **22** (1975) 161–175.
6. T. Dalenius and J. L. Hodges, Jr., "Minimum variance stratification," *J. Amer. Statist. Assoc.*, **54** (1959) 88–101.

7. M. Gonzalez, J. Ogus, G. Shapiro, and B. J. Tepping, "Standards for discussion and presentation of errors in survey and census data," *J. Amer. Statist. Assoc.*, part II, **70** (1975) 1–23.
8. L. A. Goodman, "Snowball sampling," *Ann. Math. Statist.*, **32** (1961) 148–170.
9. B. G. Greenberg, A. A. Abul-Ela, W. R. Simmons, and D. G. Horvitz, "The unrelated question randomized response model: theoretical framework," *J. Amer. Statist. Assoc.*, **64** (1969) 520–539.
10. M. H. Hansen and W. N. Hurwitz, "The problem of nonresponse in sample surveys," *J. Amer. Statist. Assoc.*, **41** (1946) 517–529.
11. H. O. Hartley, "Discussion of paper by F. Yates," *J. Roy. Statist. Soc.*, **109** (1946) 37.
12. ———, "Multiple frame methodology and selected applications," *Sankhyā*, **C36** (1974) 99–118.
13. D. G. Horvitz and D. J. Thompson, "A generalization of sampling without replacement from a finite universe," *J. Amer. Statist. Assoc.*, **47.** (1952) 663–685.
14. H. F. Huddleston, P. L. Claypool, and R. R. Hocking, "Optimal sample allocation to strata using convex programming," *Appl. Statist.*, **19** (1970) 273–278.
15. *International Statistical Review*, vol. 44. no. 2 (1976).
16. N. Keyfitz, "Sampling with probability proportional to size; adjustment for changes in probabilities," *J. Amer. Statist. Assoc.*, **46** (1951) 105–109.
17. D. B. Lahiri, "A method for sample selection providing unbiased ratio estimates," *Int. Statist. Inst. Bull.*, **33**, 2 (1951) 133–140.
18. ———, *NSS instructions to field workers*, 1952.
19. J. Lanke, "On the degree of protection in randomized interviews," *Int. Statist. Rev.*, **44** (1976) 197–203.
20. W. G. Madow, "On the limiting distributions of estimates based on samples from finite universes," *Ann. Math. Statist.*, **19** (1948) 535–545.
21. P. C. Mahalanobis, "Recent experiments in statistical sampling in the Indian Statistical Institute," *J. Roy. Statist. Soc.*, **109** (1946) 325–370.
22. P. J. McCarthy, "Replication: an approach to the analysis of data from complex surveys," *NCHS Report*. Ser. 2, no. 14, U.S. Dept. of HEW, Washington, 1966.
23. M. N. Murthy, "Ordered and unordered estimators in sampling without replacement," *Sankhyā*, **18** (1957) 379–390.
24. G. Nathan, "Approximate tests of independence in contingency tables from complex stratified samples," *NCHS Vital and Health Statist.*, Ser. 2, no. 53, Washington, 1973.
25. R. L. Plackett and P. J. Burman, "The design of optimum multifactorial experiments," *Biometrika*, **33** (1946) 305–325.

26. A. N. Politz and W. R. Simmons, "An attempt to get the "not at homes" into the sample without callbacks," *J. Amer. Statist. Assoc.*, **44** (1949) 9–31, and **45** (1950) 136–137.
27. D. Raghavarao and W. T. Federer, "Application of BIB designs as an alternative to the randomized response method in surveys," Paper no. BU-490-M, *Biometrics Unit Mimeo Series*, Dept. of Plant Breeding and Biometry, Cornell University, 1973.
28. D. Raj, "On the method of overlapping maps in sample surveys," *Sankhyā*, **17** (1956) 89–98.
29. ———, "Some remarks on a simple procedure of sampling without replacement," *J. Amer. Statist. Assoc.*, **61** (1966) 391–396.
30. J. N. K. Rao, "On the foundations of survey sampling," in *A Survey of Statistical Design and Linear Models*, J. N. Srivastava, ed., North-Holland, London, 1976, pp. 489–505.
31. J. N. K. Rao and D. L. Bayless, "An empirical study of the stabilities of estimators and variance estimators in unequal probability sampling of two units per stratum," *J. Amer. Statist. Assoc.*, **64** (1969) 540–559.
32. J. N. K. Rao, H. O. Hartley, and W. G. Cochran, "On a simple procedure of unequal probability sampling without replacement," *J. Roy. Statist. Soc.*, **B24** (1962) 482–491.
33. B. Rosén, "Asymptotic theory for Des Raj's estimator, I and II," *Scand. J. Statist.*, **1** (1974) 71–83, 135–144.
34. M. R. Sampford, "On sampling without replacement with unequal probabilities of selection," *Biometrika*, **54** (1967) 499–513.
35. H. Stenlund and A. Westlund, "A Monte Carlo study of simple random sampling from a finite population," *Scand. J. Statist.*, **2** (1975) 106–108.
36. F. F. Stephan, "Three extensions of a sample survey technique: hybrid, nexus and graduated sampling," in *New Developments in Survey Sampling*, N. L. Johnson and H. Smith, eds., Wiley, New York, 1969.
37. United Nations Statistical Office, "Recommendations for the preparation of sample survey reports," *Statistical Papers*, Ser. C, no. 1, rev. 2, United Nations, New York, 1964.
38. S. L. Warner, "Randomized response: a survey technique for eliminating evasive answer bias," *J. Amer. Statist. Assoc.*, **60** (1965) 63–69.

DECISION THEORY

Bruce M. Hill

1. INTRODUCTION

In principle, decision theory is the entire body of theory that is pertinent to the choice of actions by an individual or group of individuals in order to achieve specified goals. The primary aim of such a theory is to understand the various and complex ingredients of decision problems, and thereby to aid in the actual making of decisions, from the most trifling problems that one faces in daily life, to the most sophisticated problems of government. Such an all-embracing view of the scope of decision theory is useful, apart from anything else, in so far as much of human activity can thereby be encompassed in a single mathematical framework. In this way many apparently different problems are given a conceptual unity and seen to be variants of a main theme. Of course, any theory which purports to deal with such a large slice of human activity can be at best in only a primitive state, and only very simple problems can be given completely satisfactory solutions at present. Nonetheless, the mathematical formulations of the problem, the concepts that have been developed, and the specific results already obtained, do provide, at the very least, useful guidelines even for realistically complex decision problems.

In this essay we shall first trace the emergence of the modern theory of decision making through the ideas of the major contributors, particularly J. von Neumann and O. Morgenstern, A. Wald, L. J. Savage, and B. de Finetti. After sketching their results and views, we shall discuss some of the more active areas and directions of current research, and the problems and difficulties that the existing theory must face up to, in order better to meet the exigencies of real-world decision making. Although it is intended that this essay be as objective as possible, it goes without saying that both the selection of topics and their treatment will of necessity be somewhat biased by the personal interests and tastes of the author.

We take as our starting point and frame of reference the following formulation of a decision problem. Suppose that an individual must choose one and only one act from a set D of acts; and that, according to his judgment, he has partitioned the universe, S, into states s, one and only one of which must occur (or have already occurred). If he chooses act d and if state s occurs, then he suffers (or enjoys) a consequence $c = c(s, d)$. At the time of action the true state s will ordinarily not be known with certainty, and so he is obliged to act or make a decision under uncertainty. We do not as yet assume that the consequences are monetary or even measured numerically, but only that they are conjunctions of acts and states of reality, and are thereby themselves partial descriptions of possible states of the universe. To illustrate, suppose that the acts are to take, or not to take, an umbrella to work; and that the universe is partitioned according to whether or not it rains. The consequences are the various combinations, e.g., one doesn't carry an umbrella and it rains. We shall ordinarily assume that the act chosen does not affect the state s of the universe, as for example, the carrying of an umbrella presumably does not affect the rainfall (although strictly speaking such a possibility cannot be ruled out). It should be emphasized that both the set D of acts and the partition of the universe are matters of judgment, and the skill and imagination with which these are chosen are ordinarily the most crucial aspects of the decision problem. Thus in the umbrella example, one might wish to enlarge the space of acts to include the taking of some data, perhaps a peep out the window or listening to one's favorite weather forecaster, before departing. Similarly, the partition of the universe might include the quantity and temperature of the rain, and perhaps even whether or not one will get a ride from a neighbor (which could quite easily depend upon whether one is carrying an umbrella). Thus even in this simple example we see that upon closer examination subtleties emerge. We shall return to consideration of the difficulties hinted at here. But it will be convenient to develop the theory in several stages. Although there were many important early contributions, from Pascal to Borel, undoubtedly the first and most important stage in the development of the modern theory of decision making was the pioneering treatise by von Neumann and Morgenstern [1].

2. VON NEUMANN AND MORGENSTERN UTILITY

Without quantification of consequences, and more generally of gambles that give rise to different consequences with specified probabilities, there would be little hope for a useful theory of decision making. It was in the brilliant and now classic *Theory of Games and Economic Behavior* [1], by von Neumann and Morgenstern, that this first obstacle was overcome. Curiously, they were not concerned with decision theory as such, but rather with modelling some aspects of economic behavior in terms of games.

Their key insight lay in the recognition that if one had an ordering by preference of a set of consequences, then an assertion that $C > A > B$ (read C preferred to A preferred to B), together with an assertion that A is preferred to a gamble giving rise to either B or C, each with probability 1/2, would provide a basis for asserting that the preference for A over that for B exceeds the preference for C over that for A. Furthermore, if there were a number α between 0 and 1 such that A was equally preferred to a gamble giving rise to C with probability α, or to B with probability $1 - \alpha$, then α could be used to quantify the ratio of the preference for A over B to the preference for C over B. Such a quantification of differences between preferences is precisely what is needed to obtain "distances" between consequences, much as Euclid did for distances between points of the line, and von Neumann and Morgenstern proceeded to specify axioms under which a complete quantification of utility would be possible. They viewed the quantification of utility as analogous to the by then successful quantification of temperature, and had great expectations for a scientific theory of utility.

Their original derivation of the existence of a utility function was lengthy and given in the Appendix of [1]. Here we shall give a derivation that is based upon that of von Neumann and Morgenstern together with that of L. J. Savage in his *The Foundations of Statistics* [2]. But first it is necessary to make some remarks about the interpretation of probability. Von Neumann and Morgenstern chose to regard probability as relative frequency in an appropriate infinite sequence of trials, i.e., the conventional interpretation. At the same time they recognized that if, instead, probability was interpreted as a subjective measure of degree of belief, then their argument for the quantification of utility could still be carried out, and would also yield a well-defined notion of subjective probability [1, p. 19] as a by-product. Savage [2] carried out such a program, obtaining both utility and subjective probability by the same underlying arguments. Here we shall not beg the question, but merely treat probability as a primitive concept, obeying the usual rules, which can be interpreted according to the tastes of the reader. In particular, we shall assume that the concept of a gamble $\underline{f}$, having possible consequences f_i, $i = 1, \ldots, n$, with corresponding probabilities p_i, $i = 1, \ldots, n$, with $p_i \geqslant 0$, $\sum_1^n p_i = 1$, is an operation with concrete meaning for the reader. The properties of such gambles necessary for our derivation of utility will be included in the axioms.

AXIOMS

Let F_0 be a set of elements, $\{f, g, h, \ldots\}$, called consequences. For numbers $0 \leqslant p_i \leqslant 1$, $\sum_1^n p_i = 1$, and $f_i \in F_0$, an operation,

$\underset{\sim}{f} = \sum_1^n p_i f_i$ is defined, and referred to as the gamble with consequences f_i and corresponding probabilities p_i. The set of all possible such gambles, $n = 1, 2, \ldots$, is called $F = \{\underset{\sim}{f}, \underset{\sim}{g}, \underset{\sim}{h}, \ldots\}$. A gamble, all of whose consequences are identical, is identified with that single consequence. The following axioms are assumed.

(i) There is a complete ordering of F by a relation $\underset{\sim}{f} > \underset{\sim}{g}$. That is to say, for any two elements $\underset{\sim}{f}, \underset{\sim}{g} \in F$, either $\underset{\sim}{f} > \underset{\sim}{g}$, $\underset{\sim}{f} < \underset{\sim}{g}$, or $\underset{\sim}{f} = \underset{\sim}{g}$, where $\underset{\sim}{f} < \underset{\sim}{g}$ means $\underset{\sim}{g} > \underset{\sim}{f}$. If $\underset{\sim}{f} < \underset{\sim}{g}$ and $\underset{\sim}{g} < \underset{\sim}{h}$ then $\underset{\sim}{f} < \underset{\sim}{h}$.

(ii) If $\underset{\sim}{f} = \sum_1^m p_i f_i$ and $\underset{\sim}{g} = \sum_1^n \tilde{p}_i g_i$ are two gambles which give rise to the same set of consequences with the same probabilities, i.e., generate the same probability distribution over consequences, then $\underset{\sim}{f} = \underset{\sim}{g}$ in the sense of the ordering.

(iii) If $\underset{\sim}{f} < \underset{\sim}{g}$ and $0 < \alpha < 1$, then $\underset{\sim}{f} < \alpha\underset{\sim}{f} + (1 - \alpha)\underset{\sim}{g}$; if $\underset{\sim}{f} > \underset{\sim}{g}$ and $0 < \alpha < 1$, then $\underset{\sim}{f} > \alpha\underset{\sim}{f} + (1 - \alpha)\underset{\sim}{g}$.

(iv) If $\underset{\sim}{f} < \underset{\sim}{g} < \underset{\sim}{h}$, then there exists an α, $0 < \alpha < 1$, with $\alpha\underset{\sim}{f} + (1 - \alpha)\underset{\sim}{h} < \underset{\sim}{g}$; and there also exists such an α with $\alpha\underset{\sim}{f} + (1 - \alpha)\underset{\sim}{h} > \underset{\sim}{g}$.

(v) For some $\underset{\sim}{f}$ and $\underset{\sim}{g}$, $\underset{\sim}{f} < \underset{\sim}{g}$.

(vi) If $\underset{\sim}{g} = \underset{\sim}{h}$ and $0 \le \alpha \le 1$, then $\alpha\underset{\sim}{f} + (1 - \alpha)\underset{\sim}{g} = \alpha\underset{\sim}{f} + (1 - \alpha)\underset{\sim}{h}$.

Under these axioms we shall now prove the existence of a real valued function $U(\cdot)$, defined on F_0, and such that $\underset{\sim}{f} = \sum_1^m p_i f_i \leqslant \underset{\sim}{g} = \sum_1^n \tilde{p}_i g_i$, if and only if $\sum_1^m p_i U(f_i) \leqslant \sum_1^n \tilde{p}_i U(g_i)$. Such a function is called a utility function.

THEOREM 2.1: *If* $\underset{\sim}{f} < \underset{\sim}{g}$ *and* $0 < \alpha < \beta < 1$, *then* $(1 - \alpha)\underset{\sim}{f} + \alpha\underset{\sim}{g} < (1 - \beta)\underset{\sim}{f} + \beta\underset{\sim}{g}$.

Proof: Let $\alpha = \gamma\beta$, where $0 < \gamma < 1$. By (iii), $\underset{\sim}{f} < (1 - \beta)\underset{\sim}{f} + \beta\underset{\sim}{g}$, and so again, by (iii), $(1 - \beta)\underset{\sim}{f} + \beta\underset{\sim}{g} > \gamma[(1 - \beta)\underset{\sim}{f} + \beta\underset{\sim}{g}] + (1 - \gamma)\underset{\sim}{f}$. But by (ii), $\gamma[(1 - \beta)\underset{\sim}{f} + \beta\underset{\sim}{g}] + (1 - \gamma)\underset{\sim}{f} = (1 - \alpha)\underset{\sim}{f} + \alpha\underset{\sim}{g}$, which completes the proof.

Our next theorem provides the means for attaching a unique number in $(0, 1)$ to any gamble $\underset{\sim}{h}$ such that $\underset{\sim}{f} < \underset{\sim}{h} < \underset{\sim}{g}$, where $\underset{\sim}{f}$ and $\underset{\sim}{g}$ are any fixed gambles.

THEOREM 2.2: *Let* $\underset{\sim}{f} < \underset{\sim}{g}$. *Corresponding to any* $\underset{\sim}{h}$, $\underset{\sim}{f} < \underset{\sim}{h} < \underset{\sim}{g}$, *there exists a unique number* $\alpha \equiv \alpha(\underset{\sim}{h})$, $0 < \alpha < 1$, *such that* $\underset{\sim}{h} = (1 - \alpha)\underset{\sim}{f} + \alpha\underset{\sim}{g}$.

Proof: Consider the mapping of the open unit interval, $0 < \alpha < 1$, into the space F of gambles, defined by $\underline{h}(\alpha) = (1 - \alpha)\underline{f} + \alpha \underline{g}$. By Theorem 2.1 this mapping is strictly monotonically increasing. It remains only to show that this mapping is onto the set of all gambles, $\underline{h}, \underline{f} < \underline{h} < \underline{g}$, for in this case the mapping can be inverted to yield a necessarily unique α satisfying the conditions of the Theorem. Define $L = \{\alpha > 0 : (1 - \alpha)\underline{f} + \alpha\underline{g} < \underline{h}\}$. By (iv), L is not empty. If $\alpha_2 \in L$ and if $0 < \alpha_1 < \alpha_2$, then by Theorem 2.1, also $\alpha_1 \in L$, so that L is an interval, either open at both ends, or else closed from above. Now let α^* be the least upper bound of L and suppose $\alpha^* \in L$. In this case $(1 - \alpha^*)\underline{f} + \alpha^*\underline{g} < \underline{h} < \underline{g}$, so by (iv), there exists λ, $0 < \lambda < 1$, with $(1 - \lambda)[(1 - \alpha^*)\underline{f} + \alpha^*\underline{g}] + \lambda\underline{g} < \underline{h}$. But $\alpha^* < 1$, and $(1 - \lambda)\alpha^* + \lambda = \alpha^* + \lambda(1 - \alpha^*) > \alpha^*$, which contradicts the assumption that α^* was the least upper bound of L. (Here we have used (ii) for the equivalence of $(1 - \lambda)[(1 - \alpha^*)\underline{f} + \alpha^*\underline{g}] + \lambda\underline{g}$ and $(1 - \lambda)(1 - \alpha^*)\underline{f} + [(1 - \lambda)\alpha^* + \lambda]\underline{g}$.) It follows that α^* is not in L, so L is some open interval, say $(0, c)$, where $c < 1$. Defining $K = \{\alpha < 1 : (1 - \alpha)\underline{f} + \alpha\underline{g} > \underline{h}\}$, and proceeding similarly, we conclude that K is also an open interval, say $(d, 1)$, where clearly $d \geqslant c$. But $c < d$ would imply that $\underline{h} = (1 - x)\underline{f} + x\underline{g}$ for every x in $[c, d]$, contradicting Theorem 2.1, so that in fact $c = d$. The common value $c = d$ is then the unique $\alpha(\underline{h})$ for which $(1 - \alpha)\underline{f} + \alpha\underline{g} = \underline{h}$.

We are now in a position to characterize the quantification of consequences that is the object of this section. We defined a utility function to be any function $U(\cdot)$ mapping F_0 into the real numbers in such a way that $\underline{f} = \sum_1^m p_i f_i \leqslant \underline{g} = \sum_1^n \tilde{p}_i g_i$, if and only if $\sum_1^m p_i U(f_i) \leqslant \sum_1^n \tilde{p}_i U(g_i)$. Hence preference between gambles in F is completely determined by the expectation of utility with respect to the probability distribution that defines the gamble. Clearly if $U(\cdot)$ is a utility function, then so also is $U' = cU + d$, where $c > 0$ and d are constants. Hence if there exists a utility function, and if $f < g$ are any consequences, then we can find a utility function U for which $U(f)$ and $U(g)$ are any preassigned numbers, provided $U(f) < U(g)$. We now show that in fact any two utility functions are necessarily increasing linear functions of one another.

THEOREM 2.3: *If U and U' are utility functions, then there exist numbers c and d, such that* $U' = cU + d$, $c > 0$.

Proof: Let $f_1 \leqslant f_2 \leqslant f_3$ be any three consequences, and consider the three row vectors $v_i = (1, U(f_i), U'(f_i))$, $i = 1, 2, 3$. We wish to show that the determinant of the matrix whose three rows are v_1, v_2, and v_3, is necessarily 0. If any two of the consequences are equal, then two rows of the matrix are identical, so the determinant is 0. Without loss of generality, then, we can assume $f_1 < f_2 < f_3$. By Theorem 2.2 there exists α such that $f_2 = (1 - \alpha)f_1 + \alpha f_3$. By the defining property of a utility function, $U(f_2) = (1 - \alpha)U(f_1) + \alpha U(f_3)$, and $U'(f_2) = (1 - \alpha)U'(f_1) + \alpha U'(f_3)$. Therefore $v_2 = (1 - \alpha)v_1 + \alpha v_3$, the v_i are linearly dependent, and so the determinant is 0 as claimed. Expanding the determinant then yields

$$U'(f) = cU(f) + d,$$

as desired, where

$$c = \frac{U'(f_3) - U'(f_2)}{U(f_3) - U(f_2)} > 0,$$

and

$$d = \frac{U(f_3)U'(f_2) - U(f_2)U'(f_3)}{U(f_3) - U(f_2)}.$$

Here f is an arbitrary consequence, and of course $U(f_3) - U(f_2) > 0$.

From Theorem 2.3 it follows that if two utility functions agree at any two consequences $f \neq g$, then they are identical. Now define a set of gambles G to be convex if for every $\underline{f}, \underline{g} \in G$ and P, $0 \leqslant P \leqslant 1$, $P\underline{f} + (1 - P)\underline{g} \in G$. If f and g are consequences with $f \leqslant g$, then define the interval I of gambles determined by f and g to be the set of all gambles $\underline{h}$ with $f \leqslant \underline{h} \leqslant g$. Finally define a hyper-utility V on a convex set G of gambles to be a real-valued function of the gambles of G, such that $\underline{f} \leqslant \underline{g}$, if and only if $V(\underline{f}) \leqslant V(\underline{g})$, and such that $V(P\underline{f} + (1 - P)\underline{g}) = PV(\underline{f}) + (1 - P)V(\underline{g})$, for $0 \leqslant P \leqslant 1$. We now prove

THEOREM 2.4: *There exists a utility function.*

Proof: Let $f \leqslant \underline{h} \leqslant g$ where $f < g$. By a slight and obvious extension of Theorem 2.2, there exists a unique number, say $V(\underline{h})$, such that $0 \leqslant V(\underset{\sim}{h}) \leqslant 1$, and $\underset{\sim}{h} = [1 - V(\underset{\sim}{h})]f + V(\underset{\sim}{h})g$. Let $\underset{\sim}{h}_1$, $\underset{\sim}{h}_2$ be any gambles in the interval I determined by f and g. By (vi and ii),

$$\begin{aligned} P\underset{\sim}{h}_1 + (1 - P)\underset{\sim}{h}_2 &= P\{[1 - V(\underset{\sim}{h}_1)]f + V(\underset{\sim}{h}_1)g\} \\ &\quad + (1 - P)\{[1 - V(\underset{\sim}{h}_2)]f + V(\underset{\sim}{h}_2)g\} \\ &= \{P[1 - V(\underset{\sim}{h}_1)] + (1 - P)[1 - V(\underset{\sim}{h}_2)]\}f \\ &\quad + [PV(\underset{\sim}{h}_1) + (1 - P)V(\underset{\sim}{h}_2)]g, \end{aligned}$$

since the last expression defines the same probability distribution over the consequences. Hence $V(P\underset{\sim}{h}_1 + (1 - P)\underset{\sim}{h}_2) = PV(\underset{\sim}{h}_1) + (1 - P)V(\underset{\sim}{h}_2)$, and by Theorem 2.1, $\underset{\sim}{h}_1 \leqslant \underset{\sim}{h}_2$ if and only if $V(\underset{\sim}{h}_1) \leqslant V(\underset{\sim}{h}_2)$. Since the interval I is convex, it follows that $V(\cdot)$ defined on I is a hyper-utility.

From here on let $f < g$ be any two fixed consequences. By the same proof as given for Theorem 2.3, there is a unique hyper-utility function assigning the values 0 and 1 to f and g, respectively, on any one interval containing f and g. Let I_1 and I_2 be any two intervals containing f and g, and let V_1 and V_2 be corresponding hyper-utilities on these intervals, with $V_1(f) = V_2(f) = 0$ and $V_1(g) = V_2(g) = 1$. The intersection of I_1 and I_2 is a convex set containing f and g, so that V_1 and V_2 are identical on this intersection. Now for any gamble $\underset{\sim}{h}$ we can find an interval containing f, g, and $\underset{\sim}{h}$. Let $V(\underset{\sim}{h})$ be the common value assigned to $\underset{\sim}{h}$ by all hyper-utilities that are defined on intervals containing f, g, and $\underset{\sim}{h}$, and that assign the values 0 and 1 to f and g, respectively. Since there is always at least one such interval for any gamble $\underset{\sim}{h}$, the function V is therefore uniquely defined on the convex set F of all gambles. Let $\underset{\sim}{h}_1$ and $\underset{\sim}{h}_2$ be any two gambles and P a number, $0 \leqslant P \leqslant 1$. There exists an interval containing $\underset{\sim}{h}_1, \underset{\sim}{h}_2, f, g$ and $P\underset{\sim}{h}_1 + (1 - P)\underset{\sim}{h}_2$. In that interval the function V is a hyper-utility. Therefore $V(P\underset{\sim}{h}_1 + (1 - P)\underset{\sim}{h}_2) = PV(\underset{\sim}{h}_1) + (1 - P)V(\underset{\sim}{h}_2)$, and $\underset{\sim}{h}_1 \leqslant \underset{\sim}{h}_2$ if and only if $V(\underset{\sim}{h}_1) \leqslant V(\underset{\sim}{h}_2)$. Now let U be the restriction of V to F_0, i.e., $U(f) = V(f)$ for all $f \in F_0$, and U is not defined elsewhere. Then U is the desired utility function. For let $\underset{\sim}{f} = \sum_1^m p_i f_i$ and $\underset{\sim}{g} = \sum_1^n \tilde{p}_i g_i$ be any two gambles. Then $\underset{\sim}{f} \leqslant \underset{\sim}{g}$ if and only if $V(\underset{\sim}{f}) \leqslant V(\underset{\sim}{g})$. But $V(\underset{\sim}{f}) = \sum_1^m p_i V(f_i) =$

$\sum_1^m p_i U(f_i)$, and $V(g) = \sum_1^n \tilde{p}_i V(g_i) = \sum_1^n \tilde{p}_i U(g_i)$, so the proof is complete.

Having established the existence of a utility function, there are a number of remarks which are necessary to put this result into perspective.

Remark 1. The notion of utility in riskless situations had played a prominent role in economics prior to the work of von Neumann and Morgenstern. However, the relationship between consequences in riskless situations amounted only to stating, for any pair, which was preferred, and so, in effect, utility was only defined up to strictly increasing monotonic transformations. The analogue, with regard to temperature, would be as if one could say when an object was hotter than another, but nothing more. Another important notion, closely related to riskless utility, was that of an "indifference curve," that is to say, various collections of commodities such that an individual was indifferent as to which such collection he possessed. As von Neumann and Morgenstern pointed out, the basic assumptions underlying the existence of such indifference curves could, with only a slight extension, yield much more, namely their own utility function, uniquely defined up to increasing linear transformations. See Stigler [3] for a historical study of utility in riskless situations, and indifference curves.

Much earlier, of course, Daniel Bernoulli [4], in analyzing the St. Petersburg paradox, had been led to a notion of utility akin to that of von Neumann and Morgenstern.

Remark 2. There has been a great deal of discussion of the von Neumann and Morgenstern axioms, which are essentially the same as those we have employed. Von Neumann and Morgenstern themselves had reservations, especially as to Axiom (ii), which asserts that it is only the probability distribution of consequences that matters in determining the utility of a gamble. Another common criticism is that the complete ordering of Axiom (i) is unrealistic in so far as human preferences seem unable to be specified with the complete precision that (i) assumes. Critical discussions of the axioms appear in [1], [2], and [5].

Remark 3. The demonstration of the existence of utility given above is a modification of Savage's streamlined version of the original von Neumann-Morgenstern proof. Note that it only applies to gambles with a finite number of consequences. Savage extended the argument to apply to gambles with an infinite number of consequences. provided that the utility of these consequences is essentially bounded [2, p. 76*f*]. Blackwell and Girshick [6], assuming that the preference relation holds for gambles with a countable number of consequences, in fact prove that utility is bounded. Fishburn [7] shows that the boundedness of utility can be proved under very weak conditions. There are also convincing intuitive arguments that, apart from the type of theological questions discussed by Pascal [8], individuals, by their actions, ordinarily reveal bounded utilities [2, p. 81].

Remark 4. As our discussion of the umbrella example makes clear, there is a sense in which a consequence as dealt with here, is really itself a gamble. Thus upon further partitioning of the universe, according as to quantity and quality of the rain, the consequence (rain, no umbrella) becomes a gamble with various other more detailed consequences. But as de Finetti [9, p. 269] argues, any partitioning of the universe must be taken as provisional, and always capable of further refinement. We are thus confronted with a fundamental difficulty, namely, that of choosing an appropriate "small world" in which to make our analysis and assign our utilities. Savage [2, p. 82] offers some guidelines for the choice of such a small world, but the problem remains a vexing one which has greatly restricted the applications of decision theory.

3. FINITE TWO-PERSON ZERO-SUM GAMES

The quantification of utility was a necessary first step in formulating a general theory of decision making. Historically, the second step lay in the development of the theory of games initiated by Borel [10] and von Neumann, culminating in [1]. The games considered by von Neumann and Morgenstern are finite both as to the number of players and the number of pure strategies for each player. They include both zero-sum games, in which the sum of payments received by all players is zero, and also the economically more significant nonzero sum games, where the total output of the society is variable. For von Neumann and Morgenstern the zero-sum two-person game was but a stepping stone in the large-scale thesis they were developing. However, as will be seen in the next section, it came to occupy an especially significant position in the theory of statistical decisions as developed by A. Wald [11]. Here we shall study the properties of zero-sum two-person games as analyzed by von Neumann and Morgenstern.

The first step in their analysis is to distinguish between the extensive form of the game, by which is meant the game as it is usually conceived, unfolding step by step according to its rules; and the normal form of the game, in which each player chooses a complete strategy for playing the game, covering all eventualities arising during the play, and sufficiently detailed so that if both players have chosen their strategies prior to the beginning of play, then the conjunction of the two strategies must lead to a conclusion of the game. In other words, a player's strategy must state what his next move would be for each possible move by the other player, including chance moves (e.g., flips of a coin, rolling of dice), that may be conveniently thought of as being performed by an impartial referee. Because of the inclusion of chance moves, the conjunction of the two strategies need not lead to a unique outcome for the game, but it must lead to one of the possible forms of termination of the game. In all but the simplest of games, the number of strategies in the above sense is prohibitively large, so that the normal form of the game is primarily a device for analyzing general mathematical characteristics of games, while the extensive form is ordinarily more useful in the analysis of a particular game.

To illustrate these concepts, a game of chess in extensive form would consist of the usual sequence of moves, i.e., a choice of which player is to be white (usually a chance move), and then proceeding step by step until the game concludes in a win, loss, or draw, for white. In the normal form of chess, a strategy would first state what initial move a player would make if he turned out to be white, what response he would make to his opponent's initial move if he turned out to be black, and then in each of the cases continue to specify his next move in each possible circumstance in which he might find himself. Although finite, the number of such strategies is so enormous as to rule out any listing.

Returning to the general two-person zero-sum game, if player 1 chooses strategy a, and player 2 chooses strategy b, then the game must be played to some conclusion, and we shall assume that at such conclusion player 1 receives $K(a, b)$ from player 2, while player 2 receives $-K(a, b)$ from player 1, so that the sum of payments is 0. When chance moves are involved, $K(a, b)$ may be a random variable, so that the conjunction of the strategies a and b leads to a gamble in the sense of the last section, and consequently each player is presumably hoping for a gamble that maximizes his expected utility. Thus strictly speaking $K(a, b)$ should be interpreted as the expected utility of the gamble defined by (a, b). The goal of player 1 is, then, in some sense to maximize $K(a, \cdot)$ by an appropriate choice of the strategy a (chosen in ignorance of his opponent's strategy); while the goal of player 2 is similarly to minimize $K(\cdot, b)$ by choice of his strategy b.

The next question that von Neumann and Morgenstern consider is the precise sense in which such maximization and minimization are possible. Suppose that the strategies for player 1 are listed $1, 2, \ldots, A$, while those for player 2 are listed $1, 2, \ldots, B$. Such strategies will be called pure strategies for reasons that will emerge shortly. If, as we shall assume, $K(a, b)$ is a function known to both players, then player 1 might decide to choose a pure strategy a_0 for which $v_1 = \min_b K(a_0, b) \geqslant \min_b K(a, b)$, for any $a \neq a_0$. In this way player 1 can guarantee himself at least v_1, and no other pure strategy can guarantee him more than this. On the other hand, player 2 can choose a strategy b_0 for which $v_2 = \max_a K(a, b_0) \leqslant \max_a K(a, b)$, for any $b \neq b_0$, and thus restrict his loss to no more than v_2, with no other strategy guaranteeing that he can lose less than v_2. We have $v_1 = \max_a \min_b K(a, b)$ and $v_2 = \min_b \max_a K(a, b)$. Since for all a, b,

$$\min_b K(a, b) \leqslant K(a, b) \leqslant \max_a K(a, b),$$

$$\max_a \min_b K(a, b) \leqslant \min_b \max_a K(a, b),$$

so always $v_1 \leqslant v_2$. When $v_1 = v_2$ it is clear that there exist strategies a_0 and b_0, such that if player 1 uses a_0 he obtains at least the common $v = v_1 = v_2$, while if player 2 uses b_0 then he loses at most v, so in fact $K(a_0, b_0) = v$ is the best that either player can guarantee for himself. When $v_1 = v_2 = v$ von Neumann and Morgenstern call the game specially strictly determined, and v is often referred to as the pure value of the game. Now define a saddle point of the matrix $K(a, b)$ to be any point (a_0, b_0) such that $K(a, b_0)$ assumes its maximum at $a = a_0$, while $K(a_0, b)$ assumes its minimum at $b = b_0$. In other words, $K(a_0, b_0)$ is simultaneously maximal in its column and minimal in its row. Then

THEOREM 3.1: *The game is specially strictly determined if and only if there exists a saddle point.*

Proof: First suppose (a_0, b_0) is a saddle point. Then

$$K(a_0, b_0) = \min_b K(a_0, b) \leqslant \max_a \min_b K(a, b) = v_1,$$

$$K(a_0, b_0) = \max_a K(a, b_0) \geqslant \min_b \max_a K(a, b) = v_2,$$

so $v_2 \leqslant v_1$. But always $v_1 \leqslant v_2$ so equality holds, and the game is specially strictly determined.

To prove necessity, suppose a_0 is such that $\min_b K(a_0, b) \geqslant \min_b K(a, b)$, for every $a \neq a_0$, and that b_0 is such that $\max_a K(a, b_0) \leqslant \max_a K(a, b)$, for every $b \neq b_0$. Now if $v_1 = v_2$, then $\max_a K(a, b_0) = \min_b \max_a K(a, b) = \max_a \min_b K(a, b) = \min_b K(a_0, b)$. Hence for any a_1, $K(a_1, b_0) \leqslant \max_a K(a, b_0) = \min_b K(a_0, b) \leqslant K(a_0, b_0)$, and thus $K(a, b_0)$ achieves its maximum at $a = a_0$. Similarly, $K(a_0, b)$ achieves its minimum at $b = b_0$, and so (a_0, b_0) is a saddle point.

Saddle points need not exist, of course, nor be unique if they do exist. However, it is easy to see that any two saddle points must have the common value $v_1 = v_2 = v$. A particular class of games, called games with perfect information, always have a saddle point, and are thus specially strictly determined. Chess is such a game, but its pure value is unknown for white as player 1.

When a game is specially strictly determined the nature of rational play seems fairly clear. Player 1, try as he may, can never guarantee himself more than v, and similarly player 2 can never guarantee losing less than v. Furthermore, if either player chooses anything other than a coordinate of a saddle point, then he may do considerably worse than v. Apart from situations where one player feels that the other will not choose a coordinate of a saddle point (presumably in the hope that neither will the other), it would appear that a sensible strategy would be for each to settle for v. On the other hand, what is to be said of the case $v_1 < v_2$? Player 1 can guarantee himself at least v_1, player 2 can restrict his loss to no more than v_2, but what of values between v_1 and v_2? Can either player improve his payoff?

The answer is yes, provided that the players adopt so-called mixed strategies. A mixed strategy for player 1 is a probability distribution, $\xi(a) = \Pr\{a\} \geqslant 0$, $\sum_{a=1}^{A} \xi(a) = 1$, on the pure strategies for player 1, and a mixed strategy for player 2 is a probability distribution $\zeta(b) = \Pr\{b\} \geqslant 0$, $\sum_{b=1}^{B} \zeta(b) = 1$, on the pure strategies for player 2. A pure strategy will be identified with the mixed strategy that attaches probability 1 to that pure strategy. Now let A^* be the set of all mixed strategies for player 1, and B^* be the set of all mixed strategies for player 2. If player 1 chooses $\xi \in A^*$ and player 2 chooses $\zeta \in B^*$, then the expected utility for player 1 is

$$K^*(\xi, \zeta) = \sum_a \sum_b \xi(a)\zeta(b)K(a, b).$$

The original game has thus been replaced by a new one, called its mixed extension, in which the spaces of strategies are much richer. The striking fact, proved by von Neumann in a variety of ways, is that the extended game is always strictly determined (the adjective "specially" is dropped in order to indicate that we are playing the extended game). Thus if $v_1^* = \sup_\xi \inf_\zeta K^*(\xi, \zeta)$ and $v_2^* = \inf_\zeta \sup_\xi K^*(\xi, \zeta)$, then always $v_1^* = v_2^* = v^*$. We first prove

THEOREM 3.2: $v_1 \leqslant v_1^* \leqslant v_2^* \leqslant v_2$.

Proof:

$$K^*(\xi, \zeta) = \sum_b K^*(\xi, b)\zeta(b) \geqslant \min_b K^*(\xi, b).$$

Hence $\inf_\zeta K^*(\xi, \zeta) \geqslant \min_b K^*(\xi, b)$. Since the pure strategy b is also in B^*, it then follows that $\inf_\zeta K^*(\xi, \zeta) = \min_b K^*(\xi, b)$. Hence

$$\begin{aligned} v_1 &= \max_a \min_b K(a, b) \leqslant \sup_\xi \min_b K^*(\xi, b) \\ &= \sup_\xi \inf_\zeta K^*(\xi, \zeta) = v_1^*, \end{aligned}$$

and similarly $v_2^* \leqslant v_2$. That $v_1^* \leqslant v_2^*$ follows by the same argument as used to show $v_1 \leqslant v_2$.

We now see that when the game is specially strictly determined, i.e., $v_1 = v_2$, then it is necessarily strictly determined also, i.e., $v_1^* = v_2^*$.

By a good strategy for player 1 in the extended game we mean any ξ_0 such that $K^*(\xi_0, \zeta) \geqslant v_1^*$ for all ζ, and by a good strategy for player 2 we mean any ζ_0 such that $K^*(\xi, \zeta_0) \leqslant v_2^*$ for all ξ. The basic theorem for finite games, known as the minimax theorem, is

THEOREM 3.3: *The mixed extension of any finite game is strictly determined, i.e., $v_1^* = v_2^* = v^*$, and each player has at least one good strategy.*

Proof: A pure strategy b for player 2 can be viewed as a column vector $K_b = (K(1, b), K(2, b), \ldots, K(A, b))'$ in Euclidean space E^A of dimension A. Similarly, a mixed strategy for player 2 can be viewed as a linear combination, $\sum_{b=1}^{B} \zeta(b)K_b$, $\zeta(b) \geqslant 0$, $\sum_1^B \zeta(b) = 1$, of the vectors K_b. The set of all such linear combinations forms a closed convex set C, that is in fact the intersection of all convex sets containing K_b, $b = 1, \ldots, B$. Thus the extended game can be viewed geometrically as one in which player 1 picks a vector ξ in E^A, player 2 picks a vector $\gamma \in C$, and player 2 pays player 1 $K^*(\xi, \gamma) = (\xi, \gamma) = \sum_{a=1}^{A} \sum_{b=1}^{B} \xi(a)\zeta(b)K(a, b)$, where $\gamma = \sum_{b=1}^{B} \zeta(b)K_b$, and (ξ, γ) is the ordinary inner product in E^A. Of course ξ must have nonnegative coordinates summing to 1. To show that the extended game is strictly determined, i.e., has a value $v^* = v_1^* = v_2^*$, it suffices to show that $v_2^* \leqslant v_1^*$. If $\gamma = (\gamma_1, \ldots, \gamma_A)' \in C$, then $\max_\xi(\xi, \gamma) = \max(\gamma_1, \ldots, \gamma_A)$, and $v_2^* = \inf_{\gamma \in C} \max_\xi(\xi, \gamma) = \inf_{\gamma \in C} \max(\gamma_1, \ldots, \gamma_A)$. To prove that $v_1^* = v_2^*$ it suffices to find a ξ_0 such that for all $\gamma \in C$, $v_2^* \leqslant (\xi_0, \gamma)$, since then $v_2^* \leqslant \inf_\gamma(\xi_0, \gamma) \leqslant \sup_\xi \inf_\gamma(\xi, \gamma) = v_1^*$.

Now let T consist of all vectors $(t_1, \ldots, t_A)' \in E^A$ with $t_i < v_2^*$, $i = 1, 2, \ldots, A$. Then T is convex and its intersection with C is empty because $v_2^* = \inf_\gamma \max(\gamma_1, \ldots, \gamma_A) \leqslant \max(\gamma_1, \ldots, \gamma_A)$ for all $\gamma \in C$. In this case there exists a hyperplane $(d, x) = c$, with $(d, x) \geqslant c$ for $x \in C$ and $(d, x) \leqslant c$ for x in the closure of T [6, p. 35]. Since C is closed there exists $\gamma_0 \in C$ for which $v_2^* = \max_\xi(\xi, \gamma_0)$, and hence the maximum coordinate of γ_0 is v_2^*. It follows that $\gamma_0 - \delta_i$ is in the closure of T, where δ_i is the vector with ith coordinate 1 and all other coordinates 0. Since γ_0 is in both C and the closure of T, $(d, \gamma_0) = c \geqslant (d, \gamma_0 - \delta_i)$, so $d_i = (d, \delta_i) \geqslant 0$, $i = 1, 2, \ldots, A$, and $\sum d_i > 0$. Let $\xi_0 = (d_1/\sum d_j, d_2/\sum d_j, \ldots, d_A/\sum d_j)'$, and note $\xi_0 \in A^*$. Let $v = c/\sum d_j$. Then $(\xi_0, \gamma) \geqslant v$ for $\gamma \in C$, and $(\xi_0, t) \leqslant v$ for t in the closure of T. Since $(v_2^*, \ldots, v_2^*)'$ is in the closure of T, $v_2^* \leqslant v$, so $(\xi_0, \gamma) \geqslant v \geqslant v_2^*$ for all $\gamma \in C$, which proves $v_1^* = v_2^* = v$. Clearly ξ_0 and γ_0 are good strategies for players 1 and 2, respectively.

Just as in our discussion following Theorem 3.1, it can be argued that rational play against an intelligent opponent consists in the use of good strategies, whose existence in the extended game is guaranteed by Theorem 3.3.

To illustrate these results consider the game of Sherlock Holmes versus Professor Moriarty [1, p. 177]. Holmes desires to go from London to Dover and then to the Continent in order to escape from Moriarty, who is chasing Holmes. As the train pulls out, with Holmes aboard, he observes Moriarty on the platform, who also sees Holmes. Suppose that Holmes' pure strategies are to leave the train at the only intermediate station, Canterbury, or else to go on to Dover. Moriarty, who has hired a special train to follow Holmes, has the same two pure strategies. Assume that if they wind up at the same place, Moriarty will kill Holmes, which has a utility of 100 for Moriarty, irrespective of which place it is. If Holmes leaves at Dover and Moriarty at Canterbury, let Moriarty's utility be -50, while if the other way around, in which case Holmes may not make good his escape to the Continent, let Moriarty's utility be 0. Although it would be rather surprising if these were also Holmes' losses, let us proceed under such an assumption. With Moriarty as player 1, and with strategy $1 \equiv$ Dover for both players, the payoff matrix is $\begin{Bmatrix} 100 & 0 \\ -50 & 100 \end{Bmatrix}$. Then $v_1 = 0$, $v_2 = 100$, and the game is not specially strictly determined. But if Moriarty goes to Dover with probability 3/5, while Holmes goes to Dover with probability 2/5, then the expected utility for each is 40, irrespective of which station the other departs at, and 40 is the value of the extended game. Note that the probability that Holmes' life terminates at the end of the play is .48.

Here we have only scratched the surface of the profound theory developed by von Neumann and Morgenstern. It would be no exaggeration to say that their formulations in [1] have more or less dictated the directions which an extraordinary variety and quantity of research have taken. Indeed, in the next section we shall see how the zero-sum two-person game became the cornerstone for Wald's theory of statistical decision functions.

4. WALD AND THE THEORY OF STATISTICAL DECISION FUNCTIONS

Von Neumann and Morgenstern were primarily concerned with the zero-sum two-person game as being the simplest interesting case of the general n-person games which they hoped would initiate a new stage in the understanding of economic phenomena. Wald [11], who was aware of the Borel and von Neumann theories of games, recognized that under a certain interpretation most of what was then thought of as statistics, and much that had not been thought of, could apparently be encompassed within a single general framework, namely that of the two-person zero-sum game. He then developed a comprehensive theory of statistical decision functions that was rich enough to include the entire program of experimentation as an element of the decision problem, of which the conventional situation of an experiment giving rise to a fixed number of observations was a very special case. It will be convenient, however, to discuss Wald's work in two stages, dealing here with the case of a fixed sample size, and in Section 6 with sequential design.

To take the simplest special case, suppose that an experiment is to be performed that will give rise to n independent random variables $X_1, \ldots, X_n$, each with common distribution F. Denoting the random n-tuple by X, and its realized value after the experiment is performed by x, the data of the experiment is then x. Based upon the observed x, the statistician is then to choose an element, say d^t, from amongst a specified set D^t of terminal acts. The reason for calling such acts terminal will become clear later when we allow other acts which specify that experimentation is to continue. It is assumed that the distribution F is known to belong to a specified set Ω of distributions, and that a nonnegative function $W(F, d^t)$ is given, which measures the loss to the statistician if he chooses act d^t when F is the true distribution. Since the actual outcome x is unknown until after the experiment is performed, Wald interpreted the task of the statistician to consist in selecting a decision function $\delta(\cdot)$ mapping the space of possible observations x into D^t, or, more generally, into a space of probability distributions on D^t. Thus if x were observed, then the statistician would choose from amongst the elements of D^t according to the distribution $\delta(x)$.

Formally we have an almost exact analogue of the zero-sum two-person game. In fact Wald went so far as to interpret player 1 as Nature, whose pure strategies are distributions $F \in \Omega$, and whose mixed strategies are probability distributions ξ on Ω; player 2 was interpreted as the statistician, whose pure strategies are functions $\delta(\cdot)$, in a space Δ, mapping the observations into D^t, and whose mixed strategies, say $\delta^*(\cdot)$, are functions mapping the observations into a space of probability distributions on D^t. In the underlying game the payoff function is $K(F, d^t) = W(F, d^t)$, while in the extended game it is $K^*(\xi, \delta^*) = \int\int W^*(F, \delta^*(x))d\xi(F)\, dF(x)$, where $W^*(F, \delta^*(x)) = \int W(F, d^t)\, dP_{\delta^*(x)}(d^t)$, and $P_{\delta^*(x)}$ is the distribution on D^t that $\delta^*(x)$ generates.

As in the previous section, distributions that attach probability 1 to some element, will be identified with that element. From here on it will be convenient to drop the * notation, interpret $\delta^*(\cdot)$ as an element of Δ, and use $K(\cdot, \cdot)$ whether or not we are in the original or extended game. It is worth noting that apart from the cardinalities of the spaces of pure strategies, which are now typically infinite, we have merely a special case of the two-person zero-sum game.

When Nature chooses a pure strategy F, and the statistician chooses decision function δ, the payoff reduces to the function $K(F, \delta)$, which as a function of F for fixed δ, is called the risk function, and denoted $r(F, \delta)$. Wald uses the risk function as the basis for judging the performance of δ. The risk function permits a partial ordering of decision functions δ, wherein δ_1 is said to be at least as good as δ_2 if $r(F, \delta_1) \leqslant r(F, \delta_2)$ for all $F \in \Omega$, and δ_1 is said to be better than δ_2 if in addition the inequality is strict for at least one F. Write $\delta_1 > \delta_2$ if δ_1 is better than δ_2. If δ is a decision function and if there exists another δ_0 such that $\delta_0 > \delta$, then δ is said to be inadmissible. δ is called admissible if it is not inadmissible.

Generally speaking there seems to be little reason for employing an inadmissible decision function, provided, of course, that a better one is known and available. It would evidently be highly advantageous for the statistician to have available a class C of decision functions such that for any $\delta \notin C$, there exists a $\delta_0 \in C$ with $\delta_0 > \delta$. Such a class is called a complete class, and if one exists that is a proper subset of Δ, then the statistician can restrict his attention to that complete class. A minimal complete class, that is to say, a complete class for which no proper subset is complete, would, of course, be even more desirable. Wald proves

THEOREM 4.1: *If a minimal complete class exists, then it is the class of all admissible decision functions.*

Proof: Let C_0 denote the class of admissible decision functions, and let C_1 be a minimal complete class. Clearly C_0 is contained in

C_1. Suppose there exists $\delta_1 \in C_1$ and $\delta_1 \notin C_0$. Since δ_1 is inadmissible there exists $\delta_2 > \delta_1$. Now δ_2 cannot be in C_1, since otherwise, removing δ_1 from C_1, we would still have a complete class. Since C_1 is complete there exists $\delta_3 \in C_1$ with $\delta_3 > \delta_2 > \delta_1$. This again contradicts minimality of C_1, so in fact $C_1 = C_0$. Clearly, if the class of admissible decision functions is complete, then it must be a minimal complete class.

Now suppose that Nature uses the mixed act ξ, and the statistician uses δ. Then the average risk, using Wald's notation, is $r^*(\xi, \delta) = \int_\Omega r(F, \delta)\, d\xi(F)$. A decision function δ_0 for which $r^*(\xi, \delta_0) \leqslant r^*(\xi, \delta)$, for all $\delta \in \Delta$, is called a Bayes solution relative to the *a priori* distribution ξ. Wald also introduces the notion of a Bayes solution in the wide sense. If ξ_i, $i = 1, \ldots, n, \ldots$, is a sequence of *a priori* distributions, then δ_0 is said to be a Bayes solution in the wide sense relative to the sequence $\{\xi_i\}$, if

$$\lim_{i\to\infty} [r^*(\xi_i, \delta_0) - \inf_\delta r^*(\xi_i, \delta)] = 0,$$

where the infimum is taken over all $\delta \in \Delta$. Such a δ_0 is evidently an approximate Bayes solution insofar as for any $\epsilon > 0$, there exists N, such that for $i > N$, $r^*(\xi_i, \delta_0) - \inf_\delta r^*(\xi_i, \delta) < \epsilon$; so very little is lost by using δ_0 when i is sufficiently large.

Wald's interest in Bayes solutions lay in the fact that under modest conditions the class of Bayes solutions in the wide sense forms a complete class. Under further conditions, holding in many statistical examples, the class of Bayes solutions in the strict sense is already a complete class.

Next, paralleling the von Neumann-Morgenstern theory of games, a decision function δ_0 is said to be a minimax solution of the decision problem if $\sup_F r^*(F, \delta_0) \leqslant \sup_F r^*(F, \delta)$ for all δ; and a prior distribution ξ_0 is said to be least favorable if $\inf_\delta r^*(\xi_0, \delta) \geqslant \inf_\delta r^*(\xi, \delta)$ for all ξ. Thus δ_0 and ξ_0 are good strategies in the sense of von Neumann-Morgenstern. Wald shows that under modest conditions there exists a minimax solution and it is a Bayes solution in the wide sense. Under further, but still modest, conditions, a minimax solution is a Bayes solution relative to a least favorable prior distribution. Apart from the formulation of the general theory of decision functions, the primary achievement of Wald in [11] is to rigorously prove the above results, thus extending the von Neumann-Morgenstern theory of finite games to the case where the cardinality of the spaces of pure strategies is infinite. The proofs depend upon the introduction of appropriate metrics and various separability, measurability, and compactness assumptions. These are too detailed to discuss here, and we shall merely indicate the main results without proof.

THEOREM 4.2: *The decision problem, viewed as a zero-sum two-person game, is strictly determined, i.e.,*

$$\sup_{\xi} \inf_{\delta} r^*(\xi, \delta) = \inf_{\delta} \sup_{\xi} r^*(\xi, \delta).$$

For any a priori distribution ξ_0, there exists a decision function δ_0 such that δ_0 is a Bayes solution relative to ξ_0, i.e.,

$$r^*(\xi_0, \delta_0) = \inf_{\delta} r^*(\xi_0, \delta).$$

There exists a minimax solution and any minimax solution is a Bayes solution in the wide sense. If ξ_0 is a least favorable prior distribution, then any minimax solution is also a Bayes solution relative to ξ_0.

If ξ_0 is a least favorable prior distribution, δ_0 a minimax solution, and A is the set of all elements F of Ω for which

$$r(F, \delta_0) < \sup_{F} r(F, \delta_0),$$

then $\xi_0(A) = 0$.

Now let D_b denote the set of all decision functions δ for which $r(F, \delta)$ is a bounded function of F. Then,

The class of all Bayes solutions in the wide sense is complete relative to D_b, i.e., for any $\delta \in D_b$, there exists a Bayes solution in the wide sense, with $\delta_0 > \delta$.

Under a further compactness assumption concerning Ω, Wald shows

There exists a least favorable prior distribution, and the class of all Bayes solutions in the strict sense is complete relative to D_b.

It is to be remarked that these theorems are valid even in the sequential case to be described in Section 6. In the nonsequential case, if Ω is a closed and bounded subset of n-dimensional Cartesian space, $D^t = \Omega$, and $W(\theta, d^t)$ is continuous jointly in θ and d^t, then the conclusions of Theorem 4.2 hold. Such conditions are satisfied in many cases of point estimation.

We now prove a simple result which often makes the determination of Bayes solutions straightforward. Given the datum $X = x$, denote by ξ_x the posterior distribution of F, given $X = x$. According to Bayes Theorem this is

$$d\xi_x(F) \propto d\xi(F)\, F_G(x),$$

where $G(x) = \int_\Omega F(x)\, d\xi(F)$ is the marginal distribution of x, and $F_G(x)$ is the Radon-Nikodym derivative of F w.r.t. G, assuming $F \ll G$ for all F. Then under a joint measurability assumption,

THEOREM 4.3:

$$\begin{aligned} r^*(\xi, \delta) &= \iint W(F, \delta(x))\, d\xi(F)\, dF(x) \\ &= \int dG(x) \int W(F, \delta(x))\, d\xi_x(F). \end{aligned}$$

Proof: Since $W(F, d^t) \geqslant 0$, Fubini's Theorem holds, and allows the evaluation of the multiple integral in any order.

As a consequence of Theorem 4.3, a Bayes rule against ξ can be determined by first obtaining the posterior distribution of F, given $X = x$, i.e., ξ_x, and then choosing $\delta(x)$ so as to minimize $\int W(F, \delta(x))\, d\xi_x(F)$. In other words $\delta(x)$ is any d^t which minimizes, for that x, the expectation of $W(F, d^t)$ with respect to the posterior distribution ξ_x. Often such a d^t is relatively easy to find. Note that such Bayes rules can be taken to be pure strategies for the statistician.

Wald's theorems also suggest some simple ways to obtain a minimax solution. For example, suppose δ_0 is a Bayes solution against ξ_0, that $\sup_F r^*(F, \delta_0) \leqslant r^*(\xi_0, \delta_0)$, and that $\sup_\xi r^*(\xi, \delta) = \sup_F r^*(F, \delta)$, for all δ. Then $\inf_\delta \sup_\xi r^*(\xi, \delta) \leqslant \inf_\delta \sup_F r^*(F, \delta) \leqslant \sup_F r^*(F, \delta_0) \leqslant r^*(\xi_0, \delta_0) \leqslant \inf_\delta r^*(\xi_0, \delta) \leqslant \sup_\xi \inf_\delta r^*(\xi, \delta)$. But just as in the case of finite games, $\sup_\xi \inf_\delta r^*(\xi, \delta) \leqslant \inf_\delta \sup_\xi r^*(\xi, \delta)$, so that the statistical game has a value $r^*(\xi_0, \delta_0)$, ξ_0 is a least favorable prior distribution, and δ_0 is a minimax solution. In some cases, however, δ_0 will be a Bayes rule in the wide sense, and ξ_0 will be a non-finite measure.

It is worth noting that according to Theorem 4.2, one way to search for a minimax rule is to look for a so-called equalizer rule, i.e., one that has constant risk $r^*(F, \delta) \equiv V$ for all F. In particular, an equalizer rule that is Bayes in the wide sense is a minimax rule.

Let us now look at some examples illustrating the theory.

Example 1. The random variable X is known to have either density $f_0(\cdot)$ or density $f_1(\cdot)$, relative to Lebesgue measure. Let $D^t = \{a_0, a_1\}$, where a_i can be interpreted as an act appropriate if $f_i(\cdot)$ were known to be the true density. Let $W(f_0, a_0) = W(f_1, a_1) = 0$, $W(f_0, a_1) = c_1 > 0$, and $W(f_1, a_0) = c_0 > 0$. A pure strategy for the statistician is a function $\delta(\cdot)$ with values in D^t. If we interpret a_1 as rejection of the hypothesis that f_0 is the true density, and a_0 as acceptance of f_0, then we have the decision-theoretic version of the classical problem of testing a simple null against a simple alternative hypothesis. A pure decision function δ is evidently equivalent to a partition of the space of observations into a region R_0 where δ takes on the value a_0, and the complementary region R_1, where δ takes on the value a_1. Thus R_1 is the so-called critical region. Corresponding to any such δ there is a risk point or vector $(r(f_0, \delta), r(f_1, \delta))' = (c_1\alpha, c_0\beta)'$, where $\alpha = \Pr\{R_1 | f_0\}$, and $\beta = \Pr\{R_0 | f_1\}$. The set of all available such risk vectors can be shown to be a closed convex set, C, which contains the points $(c_1, 0)'$, $(0, c_0)'$, and along with any $(c_1\alpha, c_0\beta)'$ also contains $(c_1[1 - \alpha], c_0[1 - \beta])$. We assume the origin is not available, since otherwise the experiment would be completely decisive as to which hypothesis was true. It is almost obvious that the risk points corresponding to the admissible decision rules are a subset of the southwest boundary of the set, i.e., those on the lower boundary curve joining $(0, c_0)'$

and $(c_1, 0)'$, but not on any vertical or horizontal line segment of that boundary. Now let H_i be the hypothesis that f_i is the true density, and let $\xi_i = \Pr\{H_i\}$, $i = 0, 1$, so that $\xi = (\xi_0, \xi_1)'$ is an *a priori* distribution. If δ is a decision function with risk vector $(c_1\alpha, c_0\beta)'$, then the average risk against ξ is $r^*(\xi, \delta) = c_1\alpha\xi_0 + c_0\beta\xi_1$. If, for $k > 0$, there exist available risk vectors $(c_1\alpha, c_0\beta)'$ lying on the line $c_1\alpha\xi_0 + c_0\beta\xi_1 = k$, then all decision functions having risk vector on this line have the same average or Bayes risk with respect to ξ, and thus these risk vectors form an indifference curve with respect to ξ. Since ξ points into the positive quadrant it is geometrically obvious that the Bayes rule against ξ can be obtained by decreasing k until a value k_0 is reached for which there is at least one available risk vector lying on the line $c_1\alpha\xi_0 + c_0\beta\xi_1 = k_0$, but such that none are available, i.e., in C, for any $k < k_0$. For this k_0, whose existence follows from the closure and convexity of C, the points common to the line and C will form a line segment, degenerating possibly to a unique point. Any decision function whose risk vector is on this line segment is a Bayes solution against ξ.

It is even more instructive, however, to obtain such Bayes decision rules directly by use of Bayes Theorem. Conditional upon the data, $X = x$, the posterior probabilities of the hypotheses are

$$\xi_i(x) = \Pr\{H_i | x\} = \xi_i f_i(x)/g(x), \qquad i = 0, 1, \qquad g(x) > 0,$$

where $g(x) = \xi_0 f_0(x) + \xi_1 f_1(x)$ is the marginal density of X. Given $X = x$, the conditional expected losses for a_1 and a_0 are then

$$r^*(\xi(x), a_1) = c_1\xi_0(x),$$
$$r^*(\xi(x), a_0) = c_0\xi_1(x),$$

where $\xi(x) = (\xi_0(x), \xi_1(x))'$. Hence, given $X = x$, the Bayes rule against ξ is to choose a_1 if $c_1\xi_0(x) < c_0\xi_1(x)$, to choose a_0 if the inequality is reversed, and to choose according to any rule whatsoever if equality holds. The condition $c_1\xi_0(x) < c_0\xi_1(x)$ is equivalent to $f_1(x)/f_0(x) > c_1\xi_0/c_0\xi_1$, where we have neglected situations for which $g(x) = 0$. We have thus been led to the well-known likelihood ratio test, which rejects H_0 if the likelihood ratio $f_1(x)/f_0(x)$ exceeds the constant $c_1\xi_0/c_0\xi_1$. According to the Neyman-Pearson Lemma

[12, p. 65] such tests are optimal in the sense of minimizing β, the so-called type 2 error, for any given type 1 error, α. In this context the Neyman-Pearson Lemma is essentially equivalent to the admissibility of the Bayes tests, and the Neyman-Pearson Lemma can thus be viewed as a corollary of Wald's Theorems. It follows from our earlier discussion that the risk point corresponding to any such likelihood ratio test lies on the southwest boundary of C. Finally, minimax solutions are decision functions whose risk points are such that $c_1\alpha = c_0\beta$.

It may be noted that while everyone agrees that only the points on the admissible portion of the southwest boundary should be considered (they form a minimal complete class), the Wald theory singles out the minimax point as of special interest, while the conventional Neyman-Pearson approach would be to choose a point with a specified α and β that seems appropriate on some (usually unspecified) grounds. The subjective Bayesian approach, which will be discussed further in the next section, would be to choose ξ as an expression of opinion about the relative credibility of H_0 versus H_1, and then use the Bayes procedure against that ξ. Since all three approaches use likelihood ratio tests, the only differences in this example concern the method by which the cutoff point is chosen. For the subjective Bayesian, as we have seen, the cutoff point for the test is $c_1\xi_0/c_0\xi_1$; for the minimax approach it is such that $c_1\alpha = c_0\beta$; for the Neyman-Pearson approach it is whatever yields the desired α or β.

Example 2. Let X be a random variable having the binomial distribution based upon n trials with "unknown" probability p of success, i.e.,

$$\Pr\{X = j\} = \binom{n}{j} p^j(1-p)^{n-j}, \qquad j = 0, 1, \ldots, n.$$

Let $D^t = \{s : 0 \leqslant s \leqslant 1\}$, and $W(p, s) = (p - s)^2$. If $f(\cdot)$ is any *a priori* density function for p, relative to Lebesgue measure, then the corresponding Bayes solution against $f(\cdot)$ is

$$\delta(x) = E\{p \mid X = x\} = \frac{\int_0^1 f(s)s^{x+1}(1-s)^{n-x}\,ds}{\int_0^1 f(s)s^x(1-s)^{n-x}\,ds},$$

for $x = 0, 1, \ldots, n$. In other words, if $X = x$ is the data, then $\delta(x)$ minimizes the posterior expected loss, i.e., it minimizes $E[(s - p)^2 \mid X = x]$, over all $s \in D^t$, where p has the posterior density

$$\frac{f(p)p^x(1 - p)^{n-x}}{\int_0^1 f(p)p^x(1 - p)^{n-x}\,dp}.$$

Now take $f(\cdot)$ to be of the symmetric beta form, i.e., $f(p) = [\Gamma(2\alpha)/\Gamma^2(\alpha)]p^{\alpha-1}(1 - p)^{\alpha-1}$, $0 \leqslant p \leqslant 1$, where $\alpha > 0$, and $\Gamma(\cdot)$ is the gamma function. In this case the Bayes estimate, given $X = x$, is $\delta_\alpha(x) \equiv E[p \mid X = x] = (\alpha + x)/(n + 2\alpha)$. Note that for the improper case $\alpha = 0$ this estimate becomes the conventional one, x/n, which is in fact the maximum-likelihood estimate. The risk function for the estimator $\delta_\alpha(X) = (\alpha + X)/(n + 2\alpha)$ is $r^*(p, \delta_\alpha) = E[(\delta_\alpha(X) - p)^2] = \mathrm{Var}(\delta_\alpha(X)) + [E(\delta_\alpha(X)) - p]^2 = np(1 - p)/(n + 2\alpha)^2 + [(\alpha + np)/(n + 2\alpha) - p]^2 = [np(1 - p) + \alpha^2(1 - 2p)^2]/(n + 2\alpha)^2$.

To obtain a minimax solution we look for an equalizer rule, i.e., an α such that $r^*(p, \delta_\alpha)$ is constant for $0 \leqslant p \leqslant 1$. When $\alpha = \sqrt{n}/2$ this is easily seen to be the case, so that $\delta_{\sqrt{n}/2}$ is a minimax decision function.

Next it is interesting to compare the risk functions for the various δ_α, $\alpha \geqslant 0$, all of which are admissible. For $\alpha = 0$, the risk function is simply $p(1 - p)/n$, having a maximum of $1/4n$ at $p = \frac{1}{2}$, and the value 0 at $p = 0$ or 1. On the other hand, if $\alpha > 0$, then $r^*(0, \delta_\alpha) = r^*(1, \delta_\alpha) > 0$, while $r^*(\frac{1}{2}, \delta_\alpha) = [n/4(n + 2\alpha)^2] < 1/4n$. Thus the risk function for δ_α, $\alpha > 0$, will be below that of δ_0 in some symmetric interval about $p = \frac{1}{2}$, and will be above on the complement of that interval. Observing that δ_0 is in fact the Bayes rule against the improper prior density $p^{-1}(1 - p)^{-1}$, and that this prior distribution emphasizes the extremes, provides further insight into such behavior. In any case we see that the conventional estimator X/n is most appropriate when p is near 0 or 1, while if p is near $\frac{1}{2}$ it is possible to improve upon δ_0 by choosing some $\alpha > 0$.

Example 3. Let $X_i \sim N(\mu_i, 1)$, $i = 1, \ldots, n$, independently. Let $D^t = E^n$, n-dimensional Euclidean space, and $W(\boldsymbol{\mu}, \mathbf{x}) = \sum_1^n (\mu_i - x_i)^2$, where $\boldsymbol{\mu} = (\mu_1, \ldots, \mu_n)'$ and $\mathbf{x} = (x_1, \ldots, x_n)'$ are vectors in E^n. If $\mathbf{X} = (X_1, \ldots, X_n)'$ then the risk function for $\mathbf{X}$ is

$$r^*(\boldsymbol{\mu}, \mathbf{X}) = E\left[\sum_1^n (\mu_i - X_i)^2\right] = n,$$

so $\mathbf{X}$ is an equalizer rule. Since it is also a Bayes rule in the wide sense, $\mathbf{X}$ is a minimax rule. However, for $n \geqslant 3$, Stein [13] showed that $\mathbf{X}$ is inadmissible. Thus not all minimax rules are admissible, and in this case the Wald theory does not suggest any answer, apart perhaps from those obtained by restricting the class of decision functions in some appropriate way. Because $\mathbf{X}$ is both the maximum-likelihood and Gauss-Markoff estimate of $\boldsymbol{\mu}$, and was generally considered to be almost the uniquely natural estimator, Stein's result caused a flurry of activity that has continued to the present time. We shall return to this example in Section 5.

Now let us consider some questions as to the suitability, for statistical decision theory, of Wald's extended von Neumann-Morgenstern approach to the zero-sum two-person game. The latter was predicated on the notion that both players were rational, each seeking to maximize his expected profit, or more generally, expected utility. It does not appear to be commonly accepted that Nature plays any such game, although the possibility cannot be ruled out. Wald, of course, was aware of such an objection, but reasoned that a minimax rule, designed to be optimal against an intelligent opponent, should not fare too badly against a neutral Nature. In any case, he felt that the mathematics of minimax theory was of interest in its own right, and played an essential role in deriving his important results on complete classes of decision functions.

Often the minimax solution is criticized as being unduly pessimistic, which it surely is, if $W(F, d^t)$ is interpreted as negative income. However, as argued by L. J. Savage [2, p. 163 ff.], this objection is not quite so telling if the minimax theory is applied to what Savage calls loss, i.e., to the weight function

$$W_1(F, d^t) \equiv W(F, d^t) - \inf_d W(F, d),$$

so that W_1 measures, so to speak, one's regret at choosing d^t as compared to using the optimal terminal decision if one had known F. Thus minimizing the maximum loss, i.e., ordinary minimax theory, would lead one to avoid any act d with a huge $\sup_F W(F, d) = W(\hat{F}, d)$, say, no matter how unlikely it might seem that such $\hat{F}$ would actually occur, while the Savage modification tends to make one avoid such acts only when there is something much better available, i.e., when $\inf_d W(\hat{F}, d)$ is sufficiently small. However, since in many examples, $\inf_d W(F, d)$ is nearly 0, it seems fair to criticize the Savage modification as unduly pessimistic also. Although, Savage, of course, did not support the minimax principle, he did nonetheless suggest conditions under which it might be of value, for example, as a compromise procedure acceptable to a group of decision-makers [2, p. 172].

However, the basic difference between the approaches of Wald and Savage arose from their different interpretations of probability. Wald regarded Nature's pure strategies, i.e., probability distributions F, as physically meaningful, having an objective reality in much the same sense as the mass of an object has an objective reality. Nature's mixed strategies, i.e., prior distributions ξ on Ω, he viewed as either nonexistent, or at any rate, typically unknown. In such situations a minimax rule, at least when admissible, seemed to Wald to be an objective way of choosing a solution having desirable risk properties. For Savage, on the other hand, all probability distributions were subjective expressions of belief, and he made little, if any, distinction between Nature's pure and mixed strategies. For him the essential inputs to decision problems consisted of utility functions and prior distributions ξ on Ω, where ξ characterized the decision-maker's subjective knowledge, based upon all previous experience, about the parameters. In this case the optimal statistical decision rule is to choose a terminal act that maximizes one's personal posterior expected utility.

5. SUBJECTIVE BAYESIAN DECISION THEORY

We have seen the important role that Bayes decision functions play in the theory of Wald. Under modest assumptions they constitute a complete class, and minimax rules are Bayes against some least favorable prior distribution. However, Wald's interpretation of Nature's prior distribution ξ was essentially that of von Neumann and Morgenstern, i.e., as the strategy of an intelligent opponent in a competitive zero-sum game. Whatever the wisdom of such an interpretation may be, it does not appear that the vision of a malevolent Nature seeking to maximize its expected payoff is altogether compelling. Furthermore, another interpretation of ξ has been available for some time now, namely that ξ merely represents one's personal beliefs as to the probabilities of the various $F \in \Omega$. From this point of view probabilities have no objective meaning, and are simply human devices to aid in inference and decision-making. Rigorous formulations of subjective probabilities have been given by Savage [2] and de Finetti [9]. That such a formulation was possible had in fact already been recognized by von Neumann and Morgenstern [1, page 19]. Indeed, if one considers some well-defined event E, say, rain tomorrow, and if one regards the gamble in which he receives the stake S if E occurs and nothing otherwise as equally preferable with the sure stake pS, then this can be used to define p as the subjective probability of E. If caution is taken with regard to avoiding large stakes S, for which utility complications occur, this procedure consists in little more than an extension of the von Neumann-Morgenstern analysis of preference amongst gambles to include gambles of this special form. Despite the close relationship between utility and subjective probability (some would say their inseparability), at the present time statisticians differ greatly as to their willingness to make use of these concepts. Some accept one but not the other, some accept both, and some accept neither. The reasons for such attitudes are undoubtedly highly complex.

If, however, one is willing to make use of both concepts, then Wald's partial ordering of decision functions becomes a complete ordering in which $\delta_1 \geqslant \delta_2$ if and only if

$$\int U(F, \delta_1(x))\, d\xi(F)\, dF(x) \geqslant \int U(F, \delta_2(x))\, d\xi(F)\, dF(x),$$

where U and ξ are the utility function and prior distribution of the decision-maker. As we saw in Section 4, after observing the data $X = x$, an optimal terminal act is any d which maximizes

$\int U(F, d)\, d\xi_x(F)$, where ξ_x denotes the posterior distribution as obtained by Bayes Theorem.

Now let us reconsider the three examples at the end of Section 4.

Example 1. Here the Wald, Neyman-Pearson, and subjective Bayesian approaches differ only in their method for choosing a cutoff point for the likelihood ratio test. For the subjective Bayesian the cutoff point is $r = c_1\xi_0/c_0\xi_1$, which is plausible on common sense grounds, insofar as it should be difficult to reject H_0 if either the cost ratio c_1/c_0, or the ratio of prior odds ξ_0/ξ_1 in favor of H_0, are large. It is crucial to note, however, that this cutoff point does not depend in any way upon the sample size. Thus if the data consists of a sample of size n from either $f_0(\cdot)$ or $f_1(\cdot)$, then the Bayes rule is to reject H_0 if the likelihood ratio $L_n \equiv \prod_{i=1}^{n} [f_1(x_i)/f_0(x_i)] > r$, and both the type 1 and type 2 errors, α_n and β_n, respectively, will be decreasing functions of n. Thus

$$\alpha_n = \Pr\{L_n > r \mid H_0\} = \Pr\{n^{-1} \ln L_n > n^{-1} \ln r \mid H_0\}.$$

Since

$$E\left\{\ln\left[\frac{f_1(X)}{f_0(X)}\right] \middle| H_0\right\} = \int \ln[f_1(x)/f_0(x)] f_0(x)\, dx < 0,$$

it follows from the law of large numbers that α_n goes to 0 as $n \to \infty$, and similarly for β_n. Within the Neyman-Pearson theory, if the same level α were chosen for different sample sizes n, then from the subjective Bayesian point of view, significance at some given level α for n large might in fact provide strong evidence *for* the null hypothesis, and at any rate, would be very different from significance at the same level α with n small. Of course, there is nothing to stop one from choosing α_n to be a decreasing function of n on intuitive grounds, but it is questionable whether this could be done satisfactorily without making use of the above type of analysis.

Example 2. Savage [2, page 203] gives an interesting objection to the minimax rule, with $\alpha = \sqrt{n}/2$, as obtained above. If $\delta_0 = X/n$, while δ denotes this minimax rule, then the ratio of their risk functions is $4p(1 - p)[1 + n^{-1/2}]^2$, which has a maximum value of $[1 + n^{-1/2}]^2$, occurring at $p = \frac{1}{2}$. Hence the risk for δ_0 is less than that for δ except when $|p - \frac{1}{2}| \leqslant \frac{1}{2}\{1 - [1 + n^{-1/2}]^{-2}\}^{1/2} \simeq (4n)^{-1/4}$. Thus unless one had evidence that p was extremely close to $\frac{1}{2}$, one would ordinarily prefer δ_0 to δ if n is large. On the other hand, as in our earlier discussion of this example, unless one had evidence that p was near 0 or 1, there would exist an $\alpha > 0$ (not depending on n) for which δ_α was preferable to both δ_0 and δ. Note that from the subjective Bayesian point of view, α, which characterizes one's prior knowledge about p, should not depend upon the sample size n, and so although the minimax rule with $\alpha = \sqrt{n}/2$ is technically a Bayes rule, it would not be employed by a strict subjective Bayesian (at least not without qualms).

Example 3. In this example the minimax estimator $\mathbf{X}$ is inadmissible. Thus the Stein estimator $\mathbf{X}_1 = \mathbf{X}[1 - (n-2)/\|\mathbf{X}\|^2]$ has everywhere smaller risk than does $\mathbf{X}$ [13]. Since the early work of Stein, still other estimators that do better than $\mathbf{X}$ have been suggested and studied, mainly from an *ad hoc* point of view [14]. There is, however, a simple and interesting Bayesian analysis of the problem. Suppose that $\mu_i \sim N(\mu, \sigma_1^2)$, independently, given μ and σ_1^2; next suppose that $X_i \sim N(\mu_i, \sigma^2)$, independently, given μ_i and σ^2; finally suppose that $(\mu, \sigma^2, \sigma_1^2)$ have a joint prior distribution in which μ is independent of (σ^2, σ_1^2), and, for convenience, that μ has the improper density that is everywhere constant. Then, given μ, σ^2, σ_1^2 and the datum $X_i = x_i$, we have

$$\mu_i \sim N(Bx_i + (1 - B)\mu, (\sigma^{-2} + \sigma_1^{-2})^{-1}),$$

where $B = [1 + \sigma^2/\sigma_1^2]^{-1}$. Given σ^2, σ_1^2, and $\bar{x} = (\sum x_i)/n$, the posterior distribution of μ is $\mu \sim N(\bar{x}, (\sigma^2 + \sigma_1^2)/n)$. Hence $E[\mu_i \mid x_i, \bar{x}, B] = Bx_i + (1 - B)\bar{x}$, while $E[\mu_i|\mathbf{x}] = \hat{B}x_i + (1 - \hat{B})\bar{x}$, where $\hat{B} = E[B|\mathbf{x}]$. This exhibits the Bayes estimate for squared error loss, i.e., the posterior expectation of μ_i, as a weighted average of x_i and $\bar{x}$. Note that $\hat{B} = 1$ yields x_i, the standard minimax estimate, while $\hat{B} = 0$ yields $\bar{x}$, the minimax estimate if it were known that the μ_i were all equal. The Stein estimator $\mathbf{X}_1$ is also a special case for which $\mu = 0$ and $\hat{B}_1 = [1 - (n-2)/\|\mathbf{X}\|^2]$, although of course $\hat{B}_1$, unlike $\hat{B}$, need not lie between 0 and 1. The value of $\hat{B}$ depends upon the precise prior distribution that is employed, and is sometimes quite sensitive to such prior information. Nonetheless such Bayes estimators are admissible (unlike $\mathbf{X}_1$), and are plausible on intuitive grounds, representing a compromise between bias and precision. Thus if σ_1^2/σ^2 were known to be large, then one should give most weight to x_i since the spread of the μ_i will be large relative to the variance σ^2 of individual measurements; on the other hand, if σ_1^2/σ^2 were known to be small, then the bias introduced by using $\bar{x}$ as an estimate of each μ_i will be swamped by the measurement errors, so that most weight should be given to $\bar{x}$ in the weighted average. At any rate the extreme cases $\hat{B} = 0$ or $\hat{B} = 1$ seem ordinarily distinctly worse than some carefully chosen compromise between bias and precision. Exact and approximate evaluations of $\hat{B}$ are obtained in Hill [15]. See also [16] and [17].

6. SEQUENTIAL DESIGN AND ANALYSIS

Wald's formulation of decision theory was sufficiently general so as to include the entire course of experimentation, as well as terminal decision making. Let $X_1, X_2, \ldots,$ be an infinite sequence of random variables that can be observed if appropriate experiments are performed. In general neither independence nor identical distributions are assumed, so that the observation of one subset of variables may represent the performance of an entirely different experiment than the observation of another subset. There is no loss in generality in assuming that at most one observation can be made on each variable, since taking $r > 1$ observations on, say, X_1, is equivalent to taking a single observation on each of, say, $X_{i_1}, \ldots, X_{i_r}$, where these variables are independently distributed like X_1. Now let the space of terminal decisions, D^t, be as before. Then the first decision that must be made is whether any observations should be taken prior to choosing a terminal decision, and if so, which. If no observations are to be made, then we need only specify which terminal decision, d^t, is to be chosen. On the other hand, if an experiment is to be performed, then it is necessary to specify a subset $\{j_1, \ldots, j_r\}$, of the positive integers, that indicates the random variables to be observed at this first stage. Sometimes it suffices to observe a single random variable at each stage, but in general it is desirable, for economic and other reasons, to allow the first and subsequent stages to consist of more than one observation. This would be the case, for example, if it were natural to perform experiments using particular designs, such as Latin squares, which require more than one observation. If $X_{j_i} = x_{j_i}$, $i = 1, \ldots, r$, are the observed data at the first stage of experimentation, then we assume that there is a non-negative cost, $c(x_{j_1}, \ldots, x_{j_r})$, which we allow to depend not only upon the indices $j_1, \ldots, j_r$, but also upon the particular values observed. Of course consideration of such costs was relevant to our initial decision as to whether experimentation should have been performed at all. After having observed the data $X_{j_i} = x_{j_i}$, $i = 1, \ldots, r$, at the first stage, the theory of optimal decision making developed in Sections 4 and 5 applies in regard to choice of a terminal decision d^t. However, once again, there is the alternative to perform further experimentation before choosing such a d^t. In this case it would be necessary to specify a new subset of variables to observe, disjoint from the previous subset. These random variables represent the second stage of experimentation; and if and when they are observed, and the appropriate cost is paid, we continue sequentially in this way, deciding at the end of each stage of experimentation whether to stop and choose a terminal decision or else to continue with a new stage. Provided that we ever do stop, then given all the data up to that stage, an optimal terminal decision is chosen according to the theory of Sections 4 and 5, and we wind up with a total loss consisting of the cost of all observations made and the loss $W(F, d^t)$ arising from our terminal decision d^t.

Now let D^e be the space of all possible decisions as to the first stage of experimentation, assuming such a stage is carried out, so that the elements of D^e are finite subsets of the positive integers. If $d^e \in D^e$ is the subset $s_1 = \{j_1, \ldots, j_r\}$, then after observations have been made on the variables with these indices, the second stage of experimentation, if it is to be carried out, will consist in choosing a subset of the space $D^e_{s_1}$, where this space contains all finite subsets of the positive integers disjoint from s_1. If this subset is $d^e_{s_1} = s_2 \in D^e_{s_1}$, then after observing the corresponding random variables, and continuing inductively, we see that after k stages of experimentation, the space of possible $(k + 1)$st stage decisions is the space

$D^e_{s_1,s_2,\ldots,s_k}$ of all subsets of the positive integers disjoint from the union of the s_i, $i = 1, \ldots, k$. Now let $D = D^t \cup D^e$, $D_{s_1,\ldots,s_k} = D^t \cup D^e_{s_1,\ldots,s_k}$, $k = 1, 2, \ldots$, and $\mathbf{x} = x_1, x_2, \ldots,$. Then a nonrandomized decision function $d(\mathbf{x}; s_1, \ldots, s_k)$ is a function defined for any $k \geqslant 0$, any $\mathbf{x} = x_1, x_2, \ldots$, and any disjoint subsets $s_1, \ldots, s_k$ of the positive integers, such that

(i) for $k = 0$, $d(\mathbf{x}) = d(0)$ is any element of D, and does not depend on $\mathbf{x}$;

(ii) $d(\mathbf{x}; s_1, \ldots, s_k)$ is any element of $D_{s_1,\ldots,s_k}$, and depends only upon those coordinates of $\mathbf{x}$ in the union of the s_i, $i = 1, \ldots, k$.

It is clear that the specification of such a nonrandomized decision function determines a prescription for carrying out experimentation, and provided that experimentation ever ceases, for choosing a terminal decision at the close. Thus if $d(0) \in D^t$, then no experimentation is performed, and a terminal decision has been selected. If $d(0) \in D^e$, say $d(0) = s_1$, then observations on the random variables with indices in s_1 are taken, and the appropriate cost is paid. If $d(\mathbf{x}; s_1) \in D^t$, then experimentation ceases, and a terminal decision has been selected; while if $d(\mathbf{x}; s_1) \in D^e_{s_1}$, then a new experiment is performed, consisting of observing the variables with indices in some subset $s_2 = d(\mathbf{x}; s_1)$ disjoint from s_1, and so on, until we first reach a stage, say k, where $d(\mathbf{x}; s_1, \ldots, s_k) \in D^t$. Ordinarily we will require that the experiment terminates with probability one. Defining randomized decision functions as appropriate mixtures of such nonrandomized decision functions, we obtain the space of decision functions from which the statistician must choose. Associated with any such decision function, say δ, will be a risk function $r(F, \delta)$ which will be composed of two parts, namely, a risk that represents the expectation of $W(F, \mathbf{d}^t)$, where $\mathbf{d}^t$ is the random terminal decision that is finally chosen under δ; and a risk that represents the expectation of the cost of experimentation under δ. Despite the much greater complexity of the space of decision functions, the problem is now formally the same as that in Section 4, and in fact, the Theorems of Wald stated there were proved by him, under suitable regularity conditions, for this general case of sequential decision functions.

To illustrate this general theory let us reconsider Example 1 of Section 4 from the sequential point of view. Historically this was the first case analyzed in detail and gave rise to Wald's sequential probability ratio test [18]. We have seen that for any fixed sample size, n, the optimal terminal decision, from almost any point of view, is to choose a_1, i.e., reject H_0, if the likelihood ratio $L_n = L(x_1, \ldots, x_n) = f_1(x_1, \ldots, x_n)/f_0(x_1, \ldots, x_n)$, is greater than some constant r, and to choose a_0 if it is less than r, where it is immaterial what is done if the likelihood ratio equals r. Under the subjective Bayesian viewpoint r must have the value $\xi_0 c_1/\xi_1 c_0$, irrespective of the sample size n. Now replace ξ_0 by ξ, write $\xi_1 = 1 - \xi$, and define $P_0(\xi) = \min\{\xi c_1, (1 - \xi)c_0\}$. Then for any ξ, $P_0(\xi)$ is the Bayes expected loss if one chooses the optimal terminal decision without observing any data, when ξ is the prior probability of H_0. Suppose that the potential observations $X_1, X_2, \ldots,$ are independent and either all have density $f_1(\cdot)$ or $f_0(\cdot)$, as before, but are now observed at a constant cost $C > 0$ per observation, so that Wald's cost function, if $X_i = x_i$, $i = 1, \ldots, n$, are observed, is $c(x_1, \ldots, x_n) = nC$, irrespective of the values x_i, $i = 1, \ldots, n$. Note that in this special case, where the observations are independent and identically distributed, it is convenient to assume that each stage consists of only one observation, and that the nth stage consists in the observation of X_n. If δ is a nonrandomized sequential decision function, then in this special case we simplify the notation and write $\delta(x_1, \ldots, x_n)$ for the decision after the nth stage of experimentation, so $\delta(x_1, \ldots, x_n)$ is either one of the terminal decisions a_i, $i = 0, 1$, or else $\delta(x_1, \ldots, x_n)$ is the decision to observe X_{n+1}. Let $r(i, \delta)$ be the Wald risk function for the decision function δ when $f_i(\cdot)$ is the true density, $i = 0, 1$. If N is the random number of observations taken when δ is used, then $r(i, \delta) = EW(f_i, \mathbf{d}^t) + CE(N|i)$, $i = 0, 1$, where $\mathbf{d}^t$ is the random terminal decision chosen under δ, and $E(N|i)$ is the expected number of observations taken when f_i is the true density and δ is used. The Bayes risk $r(\xi, \delta)$, for δ against ξ, is $r(\xi, \delta) = \xi r(0, \delta) + (1 - \xi)r(1, \delta)$, and we have $r(\xi, \delta) = \min\{P_0(\xi), P_1(\xi)\}$, where $P_1(\xi) = \inf r(\xi, \delta)$, the infimum being taken over all decision rules requiring at least one observation. The function $P_1(\xi)$, $0 \leqslant \xi \leqslant 1$, is a positive and concave function, since for any ξ_1, ξ_2 and λ in the unit interval, $P_1(\lambda\xi_1 + (1 - \lambda)\xi_2) = \inf[\lambda r(\xi_1, \delta) + (1 - \lambda)r(\xi_2, \delta)] \geqslant \lambda P_1(\xi_1) + (1 - \lambda)P_1(\xi_2) \geqslant C$, where the infimum is again taken over all rules requiring at least one observation. It is also easy to see that $P_1(\cdot)$ is continuous in the closed unit interval and that $P_1(0) = P_1(1) = C$.

Now let δ_ξ be the Bayes rule against ξ, so that $r(\xi, \delta_\xi) \leqslant r(\xi, \delta)$, for all δ. Such a rule exists in this example as a consequence of Theorem 4.2. Defining $P(\xi) = r(\xi, \delta_\xi)$, we have $P(\xi) = \min[P_0(\xi), P_1(\xi)]$. Since $P_0(0) = P_0(1) = 0 < P_1(0) = P_1(1) = C$, and since both $P_0(\cdot)$ and $P_1(\cdot)$ are continuous, it follows that either $P_0(\xi) < P_1(\xi)$, for all ξ, in which case the optimal rule requires no observations; or else, because of the concavity of $P_1(\cdot)$, there exist values, ξ', ξ'', with $0 < \xi' < \xi'' < 1$, such that $P_1(\xi) < P_0(\xi)$ if and only if $\xi' < \xi < \xi''$. Hence in this case the optimal rule requires at least one observation if and only if $\xi' < \xi < \xi''$. But after a first observation $X_1 = x_1$ is taken, ξ is transformed by Bayes rule into

$$\xi(x_1) = \xi f_0(x_1)/[\xi f_0(x_1) + (1 - \xi)f_1(x_1)],$$

so that by the same argument it now is optimal to take a further observation X_2 if and only if $\xi' < \xi(x_1) < \xi''$. Thus the optimal decision rule against ξ is to continue taking observations until the posterior probability

$$\xi(x_1, \ldots, x_n) = \xi \prod_{i=1}^{n} f_0(x_i) \Big/ \left[\xi \prod_{i=1}^{n} f_0(x_i) + (1 - \xi) \prod_{i=1}^{n} f_1(x_i)\right]$$

first falls outside the open interval (ξ', ξ''). If, when this occurs, $\xi(x_1, \ldots, x_n) \leqslant \xi'$, then the optimal terminal decision is a_1; while if $\xi(x_1, \ldots, x_n) \geqslant \xi''$, it is a_0. This completely describes the Bayes optimal decision rule against any ξ. Furthermore, the condition $\xi' < \xi(x_1, \ldots, x_n) < \xi''$ is equivalent to the condition $A < L(x_1, \ldots, x_n) < B$, where $A = \xi(1 - \xi'')/\xi''(1 - \xi)$ and $B = \xi(1 - \xi')/\xi'(1 - \xi)$. Since a first observation is taken only when $\xi' < \xi < \xi''$, it follows that $A < 1 < B$ in this case. It was in this form, as a likelihood or probability ratio test, that Wald originally proposed the procedure. From his point of view A and B were constants that could be chosen to give, at least approximately, the desired type 1 and type 2 errors, as well as to minimize the expected numbers of observations. If $N \geqslant 1$ is the random number of observations that are taken under the procedure, and if * denotes the event $\{N = n, L(X_1, \ldots, X_n) \leqslant A\}$, then

$$\begin{aligned}\Pr\{a_0 \mid H_1\} &= \sum_{n=1}^{\infty} \Pr\{a_0 \mid H_1, N = n\} \Pr\{N = n \mid H_1\} \\ &= \sum_{n=1}^{\infty} \int_* \prod_1^n f_1(x_i) \prod_1^n dx_i \\ &\leqslant A \sum_{n=1}^{\infty} \int_* \prod_1^n f_0(x_i) \prod_1^n dx_i \\ &= A \Pr\{a_0 | H_0\}.\end{aligned}$$

Similarly, $\Pr\{a_1|H_0\} \leqslant B^{-1} \Pr\{a_1|H_1\}$. Hence, the type 1 error α, is $\leqslant B^{-1}$, and the type 2 error, β, is $\leqslant A$. Since Wald showed that $\Pr\{a_0|H_i\} + \Pr\{a_1|H_i\} = 1$, $i = 0, 1$, we also have $\alpha \leqslant B^{-1}(1 - \beta)$ and $\beta \leqslant A(1 - \alpha)$, so that (α, β) lies in the portion of the unit square with $0 \leqslant \alpha \leqslant B^{-1}$, and $\beta \leqslant \min[A(1 - \alpha), 1 - B\alpha]$. Ignoring the inequality signs, and setting $B\alpha = (1 - \beta)$, $\beta = A(1 - \alpha)$, Wald obtained and examined the approximations $\alpha \approx (1 - A)/(B - A)$, $\beta \approx A(B - 1)/(B - A)$. Next, letting $Z_i = \ln[f_1(X_i)/f_0(X_i)]$, $a = \ln A$, $b = \ln B$, then $A < L(X_1, \ldots, X_n) < B$ is equivalent to $a < \sum_{i=1}^n Z_i < b$. But the Z_i are independent and identically distributed random variables, and so the problem is reduced to a standard random walk problem. Wald proved that (apart from trivial special cases) the expectation of N is finite under either hypothesis, and that

$$E\left[\sum_{i=1}^{N} Z_i | H_j\right] = E[N|H_j]E[Z_1|H_j], \qquad j = 0, 1.$$

Assuming that $\sum_1^N Z_i \approx \ln a$, when a_0 is taken, and $\sum_1^N Z_i \approx \ln b$, when a_1 is taken, Wald then obtained approximations for the expected number of observations under either hypothesis. It was later shown that amongst all tests having type 1 and type 2 errors less than or equal to α and β, respectively; and for which the expected numbers of observations is finite under either hypothesis; the Wald test with α and β as type 1 and type 2 errors minimizes the expected number of observations under either hypothesis [12, page 98]. Using modern Fourier techniques, the basic Wald results are now fairly elementary to prove [19, page 567].

Since the work of Wald, there has been an enormous amount of effort put forth to develop optimal sequential methods in more complex problems. One basic technique used is that of backward induction. When the possible number of observations has a known upper bound, say N_0, one can work backwards, asking whether, given $N_0 - 1$ observations, it would be worthwhile to take one more, and continuing in this way, until the optimal rule is determined. In practice, however, the analysis becomes quite complicated even in the bounded case, and all the more so in the unbounded case (which is usually analyzed theoretically by letting the bound N_0 go to infinity). Bellman [20] studied such problems in terms of the functional equation

$$P^*(\xi) = \min[P_0(\xi), E\{P^*(\xi(X)) + C\}],$$

which must be satisfied by the risk function $P^*(\xi)$ of an optimal procedure. Here ξ is the prior distribution (which may itself be a posterior distribution) at some stage, and $\xi(X)$ denotes the corresponding posterior distribution if one additional observation is taken. Then the functional equation merely asserts that under an optimal procedure, the continuation after an observation has been taken, must itself be an optimal procedure when the posterior distribution $\xi(X)$ is regarded as a new prior distribution. Here $P_0(\xi)$ is again the risk of the optimal decision when no observations are taken and ξ is the prior distribution.

7. AREAS OF RESEARCH

Although there has been much technical elaboration and extension of the ideas discussed in this article, the underlying concepts seem to have altered very little. In this section, I would like to indicate some of the areas that have been most actively explored in the last twenty-five years, and some of the difficulties that stand in the way of further progress.

1. For fixed sample size problems there has been a great deal of work on questions of admissibility, mainly stimulated by Example 3

of Section 4. In this example the usual estimator is minimax but inadmissible, and so from the Wald approach it was natural to search for estimators which were both better than the usual estimator (hence also minimax) and admissible. One of the techniques for searching for such estimators is to look for generalized Bayes estimators with the desired property of being superior to the usual estimator. Another technique that has been adopted in difficult problems is to restrict the class of decision procedures, find a solution in the restricted class, and then examine the properties of such a solution in the original class. For example, a restriction is sometimes made to decision procedures that are invariant under a group of transformations that seems natural for the problem. Such invariance often reflects and incorporates subjective knowledge in an apparently different way from that in which the subjective Bayesian approach incorporates such knowledge. Unfortunately, difficulties can arise, even with respect to whether the problem is first reduced by sufficiency and then by invariance, or first by invariance and then by sufficiency [21]. In complex problems, where the equalizer minimax rule is inadmissible, there seems to be no satisfactory general way to utilize the Wald theory, and at present much of the work in this area is *ad hoc*. One attempt at a systematic theory is the empirical Bayes approach, in which the parameters of interest are assumed to have a nonsubjective prior distribution, which can be estimated according to classical principles, thus reducing to a mixture problem. Decision rules can then be obtained which are sometimes asymptotically as good as the Bayes rule corresponding to the assumed "true" prior distribution. Empirical Bayes procedures seem to be mathematically equivalent to subjective Bayesian procedures, in which the prior distribution itself has unknown parameters, sometimes called hyperparameters, which are estimated from the data, by non-Bayesian, and typically, *ad hoc*, methods. In a subjective Bayesian approach such hyperparameters would also be given a prior distribution, and the corresponding Bayes rule determined. However, sometimes the data provide little, or very complicated forms of information about such hyperparameters, and the empirical Bayes procedure may be a useful approximation to the fully Bayesian procedure.

If one adopts a subjective Bayesian approach from the outset, then other types of difficulties arise. Although with an appropriate choice of prior distribution the Bayes decision rule will be admissible, in problems with many parameters it is extremely difficult to choose a prior distribution that adequately expresses one's prior knowledge. Even when this is possible, there are typically very difficult mathematical problems in obtaining the Bayes estimator or decision rule. Finally, little is known at present about the robustness of such decision rules to the choice of prior distribution, likelihood function, and utility function. There is a great need for approximate Bayes decision rules that, while perhaps sacrificing some of their optimality properties, are more robust than exact Bayes rules in complex many parameter problems [22].

2. The field of sequential design and analysis of experiments, originated by Wald in connection with the sequential probability ratio test, has been extended in many directions, and the use of sophisticated probabilistic techniques has led to simplified analysis of many problems [23], [24], [25], [26]. Nonetheless, the main obstacle to implementation of such work is that in order to solve the difficult mathematical problems that arise in determining an optimal sequential design, it is usually necessary to greatly oversimplify the structure of the problem. In fact, any solution to the optimal sequential design problem presupposes a solution to the fixed sample size decision problem, and as already indicated, under realistic degrees of complexity even fixed sample size problems become extremely difficult. Even where solutions are available, little is as yet known about their robustness. As with Bayes procedures in general, it may be wise to sacrifice some optimality in order to achieve greater robustness.

3. So far our discussion has been primarily concerned with statistical decision theory in a fairly narrow sense. There are, however, much more important issues that arise when we return to the general concepts of decision theory alluded to in the Introduction. One such issue concerns the appropriateness of small world formulations, which ignore the fact that any consequence in such a small world is ultimately itself a gamble in the larger world in which we exist. However, perhaps the most fundamental obstacle to fuller

use of decision theory arises from the fact that decisions must ordinarily be made by a group rather than an individual. What is required, therefore, is a means for pooling the different utility functions and prior distributions of its various members, in order to choose wisely a course of action for the group as a whole. Such problems of group decision making require new concepts that go far beyond those required for an individual to maximize his own expected utility. It is perhaps in this area that progress is most needed, and can be hoped for in the future development of decision theory. In a sense this would constitute a return to the larger problems that motivated von Neumann and Morgenstern when they initiated that theory.

BIBLIOGRAPHY

1. J. von Neumann and O. Morgenstern, *Theory of Games and Economic Behavior*, 2nd ed., Princeton University Press, Princeton, 1947.
2. L. J. Savage, *The Foundations of Statistics*, 2nd rev. ed., Dover, New York, 1972.
3. G. J. Stigler, "The development of utility theory," *J. Political Economy*, Part I, **58** (1950), 307–327; Part II, **58** (1950), 373–396.
4. D. Bernoulli, "Exposition of a new theory on the measurement of risk" (English translation from the Latin by L. Sommer), Econometrica, **22** (1954), 23–26.
5. R. D. Luce and H. Raiffa, *Games and Decisions*, Wiley, New York, 1957.
6. D. Blackwell and M. A. Girshick, *Theory of Games and Statistical Decisions*, Wiley, New York, 1954.
7. P. C. Fishburn, "Bounded expected utility," *Ann. Math. Statist.*, **38** (1967), 1054–1060.
8. B. Pascal, *Pensées*. Introduction and notes by L. Lafuma, Paris, Delmas, 1947.
9. B. DeFinetti, *Theory of Probability*, vol. 2, Wiley, London, 1975.
10. E. Borel, "The theory of play and integral equations with skew symmetric kernels; on games that involve chance and the skill of the players" (translated by L. J. Savage), *Econometrica*, **21** (1953), 97–124.
11. A. Wald, *Statistical Decision Functions*, Wiley, New York, 1950.
12. E. L. Lehmann, *Testing Statistical Hypotheses*, Wiley, New York, 1959.
13. C. Stein, "Inadmissibility of the usual estimator for the mean of a multivariate normal distribution," *Proc. 3rd Berkeley Symp. Math. Statist. Probability*, **1** (1956), 197–206, University of California Press, Berkeley.

14. B. Efron and C. Morris, "Stein's estimation rule and its competitors—an empirical Bayes approach," *J. Amer. Statist. Assoc.*, **68** (1973), 117–130.
15. B. M. Hill, "Exact and approximate Bayesian solutions for inference about variance components and multivariate inadmissibility," *New Developments in the Applications of Bayesian Methods*, A. Aykac and C. Brumat, eds., North-Holland, Amsterdam, 1977, 129–152.
16. ———, "On coherence, inadmissibility and inference about many parameters in the theory of least squares," *Studies in Bayesian Econometrics and Statistics in Honor of L. J. Savage*, S. E. Fienberg and A. Zellner, eds., North-Holland, Amsterdam, 1975, 555–584.
17. D. Lindley and A. Smith, "Bayes estimates for the linear model," *J. Roy. Statist. Soc.* Ser. B., **34** (1972), 1–41.
18. A. Wald, *Sequential Analysis*, Wiley, London, 1947.
19. W. Feller, *An Introduction to Probability Theory and its Applications*, vol. II, Wiley, New York, 1966.
20. R. Bellman, *Dynamic Programming*, Princeton University Press, Princeton, 1957.
21. W. J. Hall, R. A. Wijsman, and J. K. Ghosh, "The relationship between sufficiency and invariance with applications in sequential analysis," *Ann. Math. Statist.*, **36** (1965), 575–614.
22. B. Hill, "A simple general approach to inference about the tail of a distribution," *Ann. of Statist.*, **3** (1975), 1163–1174.
23. M. H. DeGroot, *Optimal Statistical Decisions*, McGraw-Hill, Hightstown, New Jersey, 1970.
24. L. E. Dubins and L. J. Savage, *How to Gamble if You Must: Inequalities for Stochastic Processes*, McGraw-Hill, Hightstown, New Jersey, 1965.
25. Y. S. Chow, H. Robbins, and D. Siegmund, *Great Expectations: The Theory of Optimal Stopping*, Houghton-Mifflin, Boston, 1970.
26. H. Chernoff, "Sequential analysis and optimal design," *SIAM*, Philadelphia, 1972.

INDEX